THE
UNIFIED THEORY
OF
PHYSICS

BY

JOSEPH M. BROWN

BASIC RESEARCH PRESS

THE
UNIFIED THEORY
OF
PHYSICS

By
Joseph M. Brown
First Edition

First Impression

ISBN: 978-0-9883180-3-8
Published By
Basic Research Press
120 East Main Street
Starkville, MS 39759
United States of America
basicresearchpress.com

SYNOPSIS

The Unified Theory of Physics is based upon the assumption of an absolute time system. The universe is filled with a gas of kinetic particles which translate and collide. Furthermore, the particles make up an ether pervading all space, make the neutrinos, make all matter, and make the photons.

The mean velocity of the ether particles is $10^{10}m/s$, the particle diameter is $10^{-34}m$, the mass is $10^{-64}kg$, the particle mass density is $10^{18}kg/m^3$, and the energy is 10^{38} joules/m^3. One cubic meter of the gaseous ether has more energy than humans on the earth could use in many centuries.

Local condensations of the ether gas produce neutrinos. Neutrinos are produced by small ($10^{-24}m$ dia) nuclear pumps. A nuclear pump is produced by a random collection of the basic particles in the following manner. A few particles become collected and aligned in the same direction and same sense. This collection forms an assembly producing a slight low pressure region which causes the assembly to grow. Eventually, with many trials, a high particle density is formed which squeezes the core particles together so that they are translating at the same velocity. The flow velocity before complete condensation, occurs at v_m, the background mean velocity. The final condensation occurs without the removal or addition of energy. Therefore, the transport velocity jumps from v_m to v_r, a 9% increase. We thus will see the parameters v_r-v_m occurring throughout this physical theory. The first place we see this is that the neutrino velocity is v_r-v_m; which, of course is c, the speed of light. The neutrino has a solid core which has a dimension of $10^{-24}m$. All neutrinos have the same angular momentum, one half of $3 \times 10^{-34} kgm^2/s^2$ Plank's constant, and each develops a thrust of 1.46 mega newtons.

Each elementary matter particles at rest consists of a single neutrino traveling in a circular path at the speed of light $c(=v_r\text{-}v_m)$. The proton has an orbital diameter which makes it have an angular momentum of $\hbar/2$.

Mass is accelerated by changing the circulars path of the neutrino to a plane spiral path which, of course, limits its velocity below the speed of light. Photons impact matter and each one is partly scattered and partly captured. The net result is that the mass grows with velocity as given by $m_v = m_o / \sqrt{1 - \beta^2}$, and orbit times encase with velocity by the factor $t_v = t_o / \sqrt{1 - \beta^2}$. Thus, we obtain the transformation equation for matter from a resting frame to a moving frame as $t_v = t_o / \sqrt{1 - \beta^2}$, $x_v = x_o \sqrt{1 - \beta^2}$, $y_v = y_o$ and $z_v = z_o$. Einstein recognized these changes in matter but he erroneously thought space itself shrinks and time dilated.

When mass is added to accelerate a charged particle of matter, the mass is captured off center. This causes the particle to oscillate as it translates. Physicists often assume the particle is changed into a wave. The path of the oscillating particle is described by the Schrödinger equation which is derived using only Newtonian mechanics.

Neutrinos are continually made randomly from the ether background. Protons (and simultaneously, hydrogen atoms), as well as anti-hydrogen atoms, are continually made from the background of neutrinos. Occasionally a single hydrogen atom will reach an anti-hydrogen atom and, thus, produce a cosmic ray. Infrequently a cluster of hydrogen atoms will encounter a cluster of anti-hydrogen atoms. The result will be a high energy cosmic ray.

All matter consists of orbiting neutrinos and anti-neutrinos. Since the formation of neutrinos and anti-neutrinos are equally likely, antimatter is as likely to be formed as matter. Thus, each piece of matter is equally likely to have a mirror image of it formed. Since anti-matter is a mirror image of matter – parity is always conserved. Physicists have erroneously assumed the mirror image of an electron is an electron, instead of the positron, and have concluded that parity is not always conserved.

When photons are emitted the angular momentum of the atom-photon system is conserved during travel of the photon. Angular momentum conservation is accomplished by the photon losing one ether particle per wavelength of travel. Most physicists ignore the angular momentum conservation and assume the photon energy loss is a Doppler shift, and that the Doppler shift is due to an expanding universe. The universe is not expanding.

ACKNOWLEDGEMENTS

I thank Brittney Patterson and McRae Hopper for typing and illustrating this book. I also thank McRae Hopper for the cover and dust jacket design.

This book is another milestone in the half century quest to develop a unified theory of physical and biological science. Along the way there have been numerous colleagues and organizations facilitating the process. I wish to acknowledge some of them now.

Stephanie Johnson and the Aerospace Corporation in El Segundo, California provided the typing from 1964-1967 which reported the initial discovery that all matter is made of mass orbiting at the speed of light. On the basis of this initial document, in 1965 the Aerospace Corporation wanted to sponsor me in obtaining a PhD in physics (my PhD is in mechanical engineering). I had no interest in additional studying of *conventional* physics.

Dr. Robert M. Wood and the McDonnell Douglas Corporation sponsored the research for the next three years from 1967 to 1970. Dr. Wood was the associate director of a large group of research scientists. He felt that the mechanical theory of the universe, which I was promoting, had promise of understanding gravity. He wanted to understand and *control* gravity for vehicle propulsion. Eventually we discovered the mechanism of gravity, but we do not know how to control it. We soon discovered the mechanism of the strong nuclear force. Later, Dr. Wood had a member of my group (Harvey Bjornlie) numerically study combinations of the basic constants of physics with the hope of finding new phenomena. This led us to studying combinations of constants which would produce the fine structure constant. This so-called numerological research produced the relation $[(v_r - v_m)/v_m]^2 = 1/137.1$, whose value is close to the value of the fine structure constant. The parameter v_r is the Maxwell-Boltzmann RMS velocity and v_m its mean velocity. Current thinking is that the speed of light is very slightly less than $v_r - v_m$.

Dr. Harmon, who had a PhD in physics from the University of California at Los Angeles, continued collaborating on developing the theory until his death in 2009. A significant discovery of Dr. Harmon was the Newtonian analysis of accelerating matter which gave the formula for mass growth with velocity. This formula, $m_v = m_o/\sqrt{1 - \beta^2}$, which he derived, using Newtonian mechanics, was postulated (i.e., assumed) by Einstein in his development of the special theory of relativity.

The next organizational support came from Mississippi State University. For two decades I was provided time for research, typing support (by Teressa Yeatman) over the years, an unlimited computing support by a large pool of talented students, and access to all the facilities of a large research university. The MSU library always had available the technical documents I needed. Throughout my two decades at MSU and the three decades since, Graham Wells, a fellow professor, provided a *sounding board* and presentation techniques for the theory.

In the early years at MSU the US Army Ballistic Missile Defense Agency became interested in the theory. They awarded MSU a contract to develop the theory. We employed Dr. Leon A. Steinert, an extremely competent mathematical physicist, to develop our concepts of electrostatics and gravity. Leon had worked with our group at the McDonnell Douglas Corporation. During the McDonnell Douglas period we had found some pertinent experimental research by C. A. Bjerknes and his son in 1876, showing that breathing spheres immersed in water could produce forces of attraction and repulsion. Dr. Steinert modeled this phenomenon with our matter particles producing the breathing sphere effect and showed the model could produce electrostatic and gravitational forces. We had solved Einstein's unified field theory problem!

During the last three decades I spent part-time managing my book stores and part-time doing physics research and writing books. *The Grand Unified Theory of Physics* was published in 2004. Sherra Weeden did the typing and illustrating for this book. *The Chemistry and Mechanics of Human Aging* was published in 2008 and was typed and illustrated by David (Tobe) Stokes. The next book, *Photons and the Elementary Particles,* was published in 2011 and was typed and illustrated by McRae Hopper. The next four books, *The Neutrino* in 2012, *Foundations of Physics* in 2012, *Physics for the Millions* in 2013, and the first edition of *The Mechanical Theory of Everything* in 2015 were typed and illustrated by Brittney Patterson. In *Physics for the Millions,* Brittney was able to make the concepts of the theory *come alive.* I thank all these employees for their work.

More generally, I thank the myriad of people mentioned in the above paragraphs and many others who have helped in the development of this theory.

Joseph M. Brown
April 18, 2020

Contents

Synopsis...IV

Acknowledgements..VII

Abstract..XV

I.   Introduction...1

II. The Postulates...3

III. The Neutrino and the Second Law of Thermodynamics...........9

    A. Introduction and Discovery of the Nuclear Pump........9

    B. The Beginning and Growth of a Neutrino.................11

    C. Sizing the Neutrino...14

    D. Overall Characteristics..17

    E. Overall View of the Flows......................................22

    F. The Subsonic Flow Region.....................................25

    G. The Compression Chamber....................................28

IV. The Stable Fundamental Matter Particles.........................33

    A. The Proton and the Fine Structure of the
    Electrostatic Field..33

    B. The Strong and Weak Nuclear Forces......................37

    C. The Electrostatic Force..43

    D. The Electron...47

    E. The Five Forces of Nature......................................51

    F. Symmetry in Physics...54

    G. The Supposed Violation of Parity............................56

V. The Hydrogen Atom, the Fine Structure Constant, Gravitation,
    and Cosmic Rays..60

Contents

A. The Hydrogen Atom and the Fine Structure
Constant..60

B. Gravitation..64

C. Origin of Cosmic Rays....................................67

VI. Relativity, the Wave Property of Matter, Magnetism, and
Comparison of Newton and Einstein Mechanics.........68

A. Electrodynamics and Relativity......................68

B. Matter Waves..78

C. Magnetism..81

D. Comparison of Newton and Einstein Mechanics..........86

i. Introduction..86

ii. Magnetic Forces and Mass Growth
with Velocity...90

iii. Theoretical Experiments to Determine
Earth's Velocity....................................95

iv. Kinematics of the Motion of Matter....................96

v. Speed of Light Relative to a Moving Frame........100

vi. Einstein's Theory....................................102

vii. Conclusions...107

VII. The Particle Property of Radiation...................109

A. Photon Structure..109

B. The Non-Expanding Universe........................115

C. The Human Cosmos.....................................121

D. How We Get Our Energy..............................130

VIII. Quantum Electrodynamics............................131

CONTENTS

A. The Schrödinger Equation..131
B. Kinematics of the Motion of a Translating Particle.....134
C. Correcting the Schrödinger Equation..........................139
D. Model of a Moving Matter Particle............................143
E. Derivation of the Corrected Schrödinger Equation......145
IX. Summary, Conclusions, and Recommendations....................149
Appendix A. The Basic Constants of the Kinetic
           Particle Theory of Physics.......................................156
Appendix B. Energy, Kinetic Energy, and Work..............................160
References..167
Index..169

Abstract

Postulated a three-dimensional space populated by a gas of inert, elastic, smooth, spherical particles all of the same mass and diameter.

1. Derived the foundations of language, mathematics, and mechanics required for the Kinetic Particle Theory of the Universe.
2. Derived classical mechanics laws.
3. Derived a counter example to the second law of thermodynamics.
4. Derived the structures of the proton, electron, and photon.
5. Derived the structures of the unstable, fundamental matter particles.
6. Derived the five forces of physics.
7. For the TCP theorum, we proved that C and P are true and T is false.
8. *Showed how to find the absolute speed of the earth and prove conclusively that Einstein's relativity is incorrect.*
9. Replaced Einstein's theory of gravitation by *breathing sphere matter*.
10. Discovered the origin of cosmic rays.
11. Derived Maxwell's equations for electrostatic charge and magnetism.
12. Derived the Schrödinger equation of quantum mechanics.
13. Disproved the big-bang theory of the universe — the universe is not expanding.
14. Discovered the fine structure of the electrostatic field. The formula for the fine structure is $\alpha=(1+m_e/m_p)^2[0.99972088\times\{(v_r-v_m)/v_m\}]^2$

where v_r is the background RMS speed and v_m is the mean speed of the ether gas.

15. Derived the four fundamental constants of physics which are:
    a. The speed of light $c = 0.99972088\,(v_r - v_m)$ where $v_r/v_m = \sqrt{3\pi/8}$ so that $v_m = 3.511353066 \times 10^9 m/s$
    b. The brutino (the basic particle of this theory) radius $= 4.052 \times 10^{-35} m$
    c. The mean free path in the ether gas $= 2.35 \times 10^{-16} m$
    d. The brutino mass $= 2.89 \times 10^{-66} kg$

I. Introduction

This book shows: 1. What the simplest physical entity is which makes up everything physical in the universe, 2. How the simplest entity makes everything.

In physical science, we identify the most basic particle, from which everything is made. We call this particle the *brutino*. The brutinos make up a gaseous ether which extends indefinitely in all directions throughout an unlimited universe. Four constants are necessary and sufficient for this theory. We show how we found the velocity of the brutinos making up this gas, how we found the mass of the brutinos, how we found their diameter, and how we found the number of brutinos per unit volume of space. We show how condensations of these particles form tornado-like assemblies which make up the elementary particles of physics, i.e. the neutrinos, electrons, protons, neutrons, and photons. We also show how the four well-known forces are produced, i.e. the electromagnetic, gravitational, weak nuclear, and the strong nuclear forces. We discovered the fifth, and most powerful, force of nature. This force is the meganewton force driving neutrinos through the dense ether. We derive all the forces as well as the common fundamental particles of physics. We derive the Schrödinger equation, using Newtonian Mechanics, which models the dynamics of the motion of matter particles and proves that quantum mechanics is a subset of classical (Newtonian) mechanics. We prove parity is always conserved. The theory is a truly unified theory of physical science.

We discovered an error in Einstein's derivation of his special theory of relativity. The essence of this error is that matter shrinks and orbital periods increase with velocity but space does not shrink nor does time dilate - as Einstein's Theory predicts.

II. The Postulates

Are there fundamental postulates from which all of physics can be derived? If so, what are these postulates? We have discovered that everything in the universe can be constructed of one type small elastic particle. Yes, just one type particle makes all other particles of matter, makes light and all other radiation, and even makes the elusive neutrinos. We call the basic particle the *brutino*. The name brutino means tiny brute, since it is very small and is the brute that makes everything. The postulates are:

1. Space is three dimensional
2. One type and size of particle makes everything
3. The particle is spherical, smooth, elastic, moves, and collides with other particles

The brutino is the smallest possible thing in the world. It is a hard, smooth, perfectly elastic sphere whose average speed is over ten times the speed of light. Most brutinos travel many diameters of distance before impacting another brutino. However, at a large number of locations in space many brutinos in continuous contact with each other produce neutrinos. Each elementay piece of matter at rest is made of a single neutrino traveling in a circular path at the speed of light. Brutinos are neither created nor destroyed, and all brutinos are exactly alike.

The brutino radius is $r_b = 4.052 \times 10^{-35}$ m, the mass $m_b = 2.89 \times 10^{-66} kg$, the average velocity $v_m = 3.511353 \times 10^9 m/s$, and the average density of the mass in the universe $\rho_0 = 4.23 \times 10^{17}$ *kg/m³* (see Appendix A). Figure 2.1 shows the brutino.

Smooth, Hard, Perfectly Elastic Sphere

$r_b = 4.052 \times 10^{-35}$ *meters*
$m_b = 2.89 \times 10^{-66}$ *kilograms*
$v_m = 3.511353 \times 10^9$ *meters per second*

Figure 2.1. The Brutino, the Ether Particle

There is nothing else in the universe except brutinos. Space without brutinos would be a vacant arena which extends indefinitely in all three directions. The gas of brutinos occupies and extends everywhere throughout space. The average number of brutinos per unit volume is $\eta_0 = \rho_0/m_b = 1.46 \times 10^{83}$ brutinos per cubic meter. The average spacing between brutinos is $s = (1/\eta_0)^{1/3} = 1.899 \times 10^{-28}$ m. The mean free path is $\ell = 1/(\sqrt{2}\,\pi d^2 \eta_0) = 2.35 \times 10^{-16}$m. The average mass density, ρ_0, throughout all space being 4.23×10^{17} *kg/m³* is a very large density but not nearly as large as the density of a proton. The gas is depicted in Figure 2.2.

Everything is made only of brutinos. All matter, radiation (radio waves, light, and x-rays), neutrinos, and even the means for transmitting forces (nuclear, electromagnetic, and gravitational) consists only of brutinos. All the atoms, the earth, stars, and living matter are made only of brutinos. The gas made up of brutinos is called the *ether*.

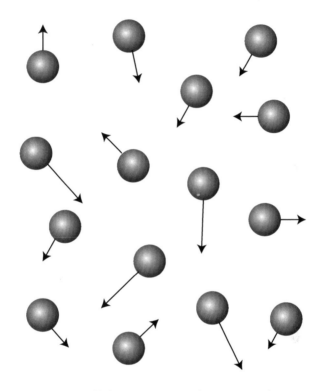

η_0=1.46×10^{83} *brutinos per cubic meter of space*
ρ_0=4.23×10^{17} *kilograms per cubic meter of space*
ℓ=2.35×10^{-16} average *meters travel distance between impacts*
v_m=3.511353×10^9 *meters per second*

$$s= \left(\frac{1}{1.46 \times 10^{83}} \right)^{1/3} =1.899 \times 10^{-28}$$ the *average number of meters between brutinos*

Figure 2.2. The Brutino Gas of Ether Particles

The most refined measurement of distance which can be made is one brutino diameter. The smallest mass measurement is the mass of one brutino. The brutino diameter controls gravitation. The brutino mass controls photon decay.

Just how basic are the four quantities which characterize the universe, i.e., the average mass density throughout the universe, the brutino mass, the brutino radius, and the average velocity of brutinos? The mass of anything, quite simply, is just the number of brutinos present times a constant. Any distance (e.g., brutino radius and mean free path) is just a matter of how many brutino diameters there are between the two end points. A velocity is simply how many brutino diameters of displacement of something occurs *while* so many brutino diameters of displacement of something else occurs. With these meager postulates, i.e., of space and the brutino characteristics, we attempt here to derive all phenomena observed in the universe.

Consider a large cubical room fixed in space with hard, perfectly elastic walls into which there are inserted a large number N of brutinos but where the combined volume of all the brutinos is very small, say less than a billion billionth of the room volume. Having such a small portion of the volume occupied by the brutinos results in a rare gas. Assume that all the brutinos have as near as possible a single speed v_{m_i}, that they are equally spaced throughout the room, and that a third of the total are moving in each of the three directions parallel to the walls. For each direction let half move in one sense and half in the opposite sense. As a result of collisions and variances from the exact requirements, the magnitudes and directions of the brutino velocities change and the most likely resulting state is known as the homogeneous state. In this state the number of brutinos in each equal volume of space, for large numbers of brutinos, or the average for large numbers of samples, is very close to being the same (percentage-wise) no matter where or when the samples are taken.

Also, for large sample sizes the number of brutinos within a given volume having directions lying within an angle dispersion defined by a given size cone is the same independent of the orientation of the cone longitudinal axis, i.e., for all directions in space and for all volumes of the space. Next, the components of the velocities parallel to a given axis of all brutinos has a Gaussian distribution

$$f(v_x) = \frac{1}{\sqrt{\pi} v_m} e^{-(v_x/v_m)^2} \tag{2.1}$$

where v_x is the component of velocity parallel to x and v_m in the particle average velocity, see Kennard, [2.1]. The distribution of the magnitude of the velocities v is the Maxwell-Boltzmann distribution

$$f(v) = \frac{4v^2}{\sqrt{\pi} v_m^3} e^{-(v/v_m)^2} \tag{2.2}$$

Again see Kennard. Figure 2.3 is a plot of $f(v/v_{mp})$ versus v/v_{mp}.

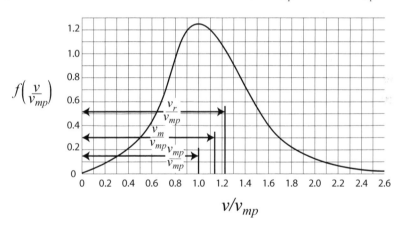

Figure 2.3 Distribution of Ether Particle Speeds

In this graph v_{mp} is the velocity with the highest ordinate, v_m is the average particle velocity, and v_r is the root mean square (RMS) velocity.

Returning to the brutinos which were all moving at the same speed, after they become distributed their energy is the same so $Nm_b v_{m_i}^2=(Nm_b)v_r^2$, where v_r is the root mean square of the particle velocities. The energy before becoming distributed is $(Nm_b)v_{m_i}^2$. Thus, since the initial energy is the same as the final energy $v_r=v_{m_i}$, i.e., the RMS velocity after becoming distributed is equal to the initial mean speed (and initial RMS speed since the two speeds are the same for the initial condition). The mean speed for the Maxwell-Boltzmann distribution is related to the RMS speed by $v_r/v_m=\sqrt{3\pi/8}$, so that the final mean speed v_m is 0.921318 times the initial mean speed. The difference in the mean speeds for the two conditions is significant. The mixing process brought about by collisions actually reduced the mean speed by 9 percent! In this distributing process the mean speed changed from $v_r (= v_{m_i})$ to v_m, i.e., the change is v_r-v_m. Later it will be shown that a process which is the reverse of this takes place in a neutrino and the mean speed of the particles in the neutrino increases from v_m to v_r. Further, the neutrino travels at the speed v_r-v_m, where v_r and v_m are the background RMS and mean speeds.

The homogeneous state for the dilute (or rare) gas of brutinos has been discussed. The homogeneous state has a uniform density of particles and has a Maxwell-Boltzmann distribution of speeds. The distribution of speeds is the same for all directions. Direct measurements have been made which verify the existence of this homogeneous state. There can be no question that this state is a stable configuration of brutino-like gas.

However, we will later show that an inhomogeneous state of the ether gas can exist. This inhomogeneous state is known as the neutrino, and it is the basis of all observables in the universe.

III. The Neutrino and the Second Law of Thermodynamics

A. Introduction and Discovery of the Nuclear Pump

The first and foremost hurdle for the kinetic particle theory of physics is how to obtain useful energy from the high energy ether gas. The problem is akin to obtaining useful energy from the air, for example, from still air in a room. Equilibrium air at room temperature has about 10,000 $joules/m^3$ of energy. Scientists have pondered this question for years, without success, of how to obtain useful energy from such a source. In fact, the scientific community has developed the second law of thermodynamics which states that useful energy can not be obtained from an isolated container of gas which is in an equilibrium state. We show that this is not true. Stable states can occur in ideal gases with long mean free paths.

In the kinetic particle theory of physics we have an ether gas with an energy of 10^{38} $joules/m^3$ — enough energy to provide for all the needs of the earth for centuries. Our research team was so certain the universe was a kinetic particle universe that we searched extensively for ways to extract energy from the ether gas. Finally, we discovered the *nuclear pump*, or, simply, the *pump*. First, however, we note that all processes in the universe conserve energy. The only processes which don't conserve energy are those which neglect some part of the process. Now, if particles are taken from a Maxwell-Boltzmann gas without changing their speeds, and aligned parallel and in the same sense, several things occur. The particle flow velocity will be v_m the background mean velocity. Their energy will be $(1/2)mv_r^2$, where m is the total mass of the particles and v_r is the RMS velocity of the background gas. For a Maxwell-Boltzmann gas $v_r/v_m = \sqrt{3\pi/8}$.

If the assembly of aligned particles described above is squeezed together without changing their energy and the particles are all touching each other so that they all have the same velocity, then that velocity must be v_r. Of course to make the particles accelerate from the flow velocity v_m to velocity v_r requires a force in the direction of the acceleration. Subsequently, we will show that this force is self-generated and, further, it is the force which organizes particles of the ether gas. It is the force which continually decreases the entropy of the universe. The force is the agent responsible for violating the second law of thermodynamics.

Recognizing the self-generating force is one facet of violating the second law of thermodynamics, but there are other significant facets. We describe these other facets by describing the origin of the neutrino.

B. THE BEGINNING AND GROWTH OF A NEUTRINO

In a gas various assemblies are made, and disintegrate continually. The beginning of a neutrino possibly could start with a few particles randomly getting aligned with all particles moving in the same direction (and sense). Such an assembly would tend to produce a vacuum region behind it. The assembly would behave as a pump and would form a *core* region in the gas. If the assembly survived for a while it would begin to *suck-in* particles from the background. Some of these particles might stay with the assembly. If the assembly randomly survives and grows some, it will suck-in more particles and, at some stage of growth, the inflow might be great enough so that the particles begin circulating about the assembly velocity vector. The measure of this circulation is the assembly angular momentum. The angular momentum can be left- or right-handed, of course. The inflow initially is radial, directed toward the center of the pump. As the flow gets closer to the pump it begins circulating, but it also begins to collide more and scatter forward and aft.

Another significant event is that the few particles (making up the core) begin to get squeezed together so that they are forced forward. This event must be in the early growth phase. Many attempts to produce assemblages may be required to produce one assemblage which is stable. Once the assemblage is stable it will continue its growth.

The output of the assembly is two fine streams - the forward stream is the solid core moving at velocity v_r and the aft stream is the semi-solid moving at velocity v_m.

The condensed exiting streams are so small that they have no measurable effect on the inflow. The particles exiting the assembly, in the fine streams, result in reducing the back-pressure upstream with the net result of incasing the inflow from all directions. The flow

1 1

reaches the speed of sound at some radius from the core. This radius will increase as the core continues to grow.

At the early stage of growth we assume that every particle which reaches the *core* will be captured then either remain with the core or be emitted forward or aft from the core. What this means is that the gas will reach sonic speed at the core. Our discussion would be more complex if only a portion of the particles were captured but we think that the end result would be the same.

The flow downstream of the sonic sphere will have streamlines directed toward the center of the core. Such a flow could reduce the thermal motion more than the flow out of a nozzle into a vacuum. If so, the sonic parameters would exist slightly upstream from the core. The sonic sphere radius then would be greater than the core *radius*. Now, gas coming into this sonic sphere from one side will be partly captured and part will exit the opposite side of the sphere. This will reduce the *back pressure* all around the sonic sphere. As a result the flow will increase and the sonic sphere radius will increase. The sonic sphere radius will continually increase until it is greater than the mean free path. When the sonic sphere radius gets much greater than the mean free path, the particles entering the sphere will swamp the inflow mechanism and experience back pressure. At some level of back pressure (which means sonic sphere radius) there will be a balance of suction flow and back pressure which will stop the neutrino growth. This will be the observed neutrino.

Consider the size of an *air neutrino* which has a core diameter to produce a sonic sphere three times the mean free path of air. Here we assume that this size would be the maximum size possible. We have

$$(3 \times \text{mean free path of standard air})^2\, \rho_{std\ air} = (r_{solid\ core})^2\, \rho_{liquid\ air} \quad (3.1)$$

Substituting numbers

$$3 \times 6 \times 10^{-8} \sqrt{1.2} = r_s \sqrt{920} \, , \quad r_s = 1.30 \times 10^{-9} m \qquad (3.2)$$

where r_s is the radius of the solid (or liquid) core. Such a small size would consist of a core with a cross section of approximately one molecule of air. This would not be adequate for solidification and for the *pump* action. If the core attempted to grow, the small mean free path length to sonic sphere radius would *swamp* the flow and destroy the would-be air neutrino.

The explanation here of the neutrino formation and growth is speculative and may, or may not, be the way neutrinos are formed. But, we know they are formed. They are stable, and we know many of their characteristics. Our explanation of their formation might be the way neutrinos are formed and with this formation process, very little random organization is required.

C. Sizing the Neutrino

As will be discussed at the beginning of the next chapter, all mater is made up of elementary matter particles. An elementary matter particle is a single orbiting neutrino. The proton is a single orbiting neutrino. The proton is formed when a neutrino having the mass of a proton gets knocked into a circular orbit. The neutrino's angular momentum of $\hbar/2$, when in the proton, is manifested as the neutrino mass moving at the speed of light in a circle whose radius has the value to make the proton have angular momentum $\hbar/2$. There is, of course, a unique value of radius which meets these requirements. Furthermore, all neutrinos produce the saame thrust, T, and that value of thrust exactly balances the proton's neutrino centrifugal force. The value of T is determined in the first section of Chapter 4. The value is given by eg. (4.3) and is

$$T = 1.49295 \times 10^6 \text{ Newtons} \tag{3.3}$$

The threust is the time rate of momentum change, which is the mass flow rate $\dot{m}$ through the neutrino times the change in velocity, which is $v_r\text{-}v_m$ ($=c$). Thus, we write the force balance equation as

$$T = \dot{m}\,c = (0.649\rho_o)(4\pi r_c^2)(0.7v_m)(c)$$
$$= (0.649 \times 4.23 \times 10^{17})(4\pi r_c^2)(0.7 \times 3.51 \times 10^9)(3 \times 10^8) \tag{3.4}$$
$$= 1.493 \times 10^6$$

Solving for r_c gives

$$r_c^2 = 5.87 \times 10^{-31}, \; r_c = 7.66 \times 10^{-16} m \tag{3.5}$$

The mean free path, l, has the value

$$l = 2.35 \times 10^{-16} m \qquad (3.6)$$

Further

$$\frac{r_c}{l} = \frac{2.66 \times 10^{-16}}{2.35 \times 10^{-16}} = 3.26 \qquad (3.7)$$

Now that we have the value of r_c we can compute the required flow rate.

$$\dot{m} = 0.649 \rho_o (4\pi r_c^2)(0.7 v_m)$$
$$= 0.649 \times 4.23 \times 10^{17} [4\pi (7.66 \times 10^{-16})^2](0.743.51 \times 10^9) \qquad (3.8)$$
$$= 4.98 \times 10^{-3} \, kg/sec$$

Incidentally we have just shown why the proton has just the mass it has. Although we used the thrust required to hold the proton in orbit to calculate the flow rate through the neutrino, the flow rate is actually produced by the mechanism of the neutrino which thus determines its thrust. However we do not know directly how to calculate the thrust hence we used the proton. What actually happens in nature is the neutrino is formed and all neutrinos produce exactly the same thrust. Neutrinos are continually knocked around but the only mass which will orbit stably is the one which will conserve angular momentum and also have angular momentum $\hbar/2$. There is only one mass which will produce angular momentum $\hbar/2$ - and that is the mass of the proton.

Concider now the angular momentum of the neutrino. Let us compute the radius r_h of a spherical volume of background gas with a periphreal velocity v_m which would have the angular momentum of the neutrino. The radius of gyration of a solid sphere is $(2/5)r_h$. Thus

$$\hbar/2 = 5.20 \times 10^{-35} = \rho_o(4/3)\pi r_h^3(2/5)r_h \times v_m$$
$$= (4.23 \times 10^{17})(4/3)\pi r_h^3(2/5)r_h \times 3.51 \times 10^9 \qquad (3.9)$$

From this

$$r_h^4 = 8.86 \times 10^{-62}, \ r_h = 5.46 \times 10^{-16} m \tag{3.10}$$

The radius of this sphere to the sonic sphere radius is

$$r_h/r_c = 5.46 \times 10^{-14}/(7.66 \times 10^{-16}) = 0.733 \tag{3.11}$$

This value of r_h seems reasonable.

D. OVERALL CHARACTERISTICS

Complete condensations of the brutinos occur randomly, but continually, and produce neutrinos. The neutrino takes in mass at the rate[1] of 10^{-2} *kg/s*. The neutrino is defined by its sonic sphere, which is an almost spherical surface at which brutinos flow into the neutrino at their speed of sound. The radius of this sphere is approximately $10^{-15}m$. The neutrinos form with masses which vary by several orders of magnitude, with right- and left-handed twist. They expel the incoming mass out of two fine streams, one directed forward at velocity v_r and one directed aft at velocity v_m. The neutrino develops a thrust of 1.43 meganewtons. The neutrino translates at the velocity $v_r - v_m$, where $v_r = \sqrt{3\pi/8}\ v_m$. It produces power of $1.43\times10^6\times3\times10^8 = 4.29\times10^{14}$ *watts*. Figure 3.1 shows the neutrino.

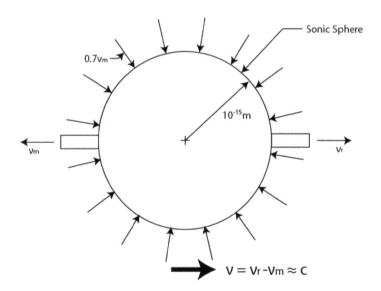

Figure 3.1 The Neutrino

The neutrino is a flow pattern in the gaseous ether which pervades the universe. The flow pattern is produced by a microrocket

1   This rate is determined from the analysis of the proton.

pump. The microrocket pump is produced by the random flows of the ether gas. The microrocket pump translates at the velocity $v_r - v_m$. The pump is cylindrical in shape and consists of a completely dense forward section and a semi-dense aft section. The pump develops a thrust of 1.43 meganewtons. Outside the microrocket pump there is the spherical microgas compressor. This section receives ether gas at its almost spherical surface (with a radius of $10^{-15}m$) at sonic speed, and compresses the gas by aligning the incoming particles, then compressing the gas to near solid, turning the particles to produce angular momentum, and turning them again to direct them forward and aft of the direction of the neutrino velocity. There also is the third section of the neutrino which consists of the subsonic flow which reaches sonic conditions just at the location where the particles enter the compression chamber. This third section is the subsonic flow region and it extends from the outside of the compression chamber indefinitely in the outward direction.

Figure 3.2 is a sketch showing the principal characteristics of the neutrino. The subsonic region is defined by

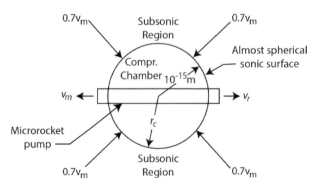

Figure 3.2. The Neutrino

the almost spherical surface which has a radius, r_c, approximately $10^{-15}m$. The gas flows into the sonic surface reaching the local speed of sound $(0.7v_m)$ at the surface. In the compression chamber the gas initially flows slowly but since the mean free path of the gas is near the radius of the compression chamber, particles coming from the opposite surface of the compression chamber impact the incoming particles less. As a result, the flow *goes critical* and almost solidifies before reaching the core. The gas then turns to circulate around the diameter of the sphere which diameter is parallel to the neutrino velocity. The gas then turns to travel parallel to the neutrino velocity approximately half the mass forward and half aft. In all this motion inside the compression chamber as well as in the subsonic region, the particle speeds do not change; only their directions change. The gas particles now all aligned parallel to the neutrino velocity are forced into the core, i.e., the microrocket pump. As half the particles are pressed into one side of the microrocket pump and half are pressed into the other side, one of the sides gets solidified and begins moving away from the microrocket pump center at velocity v_r. Meanwhile, in the other side of the microrocket pump, the particles are not able to solidify and, as a result, they move away from the microrocket pump center at velocity v_m. The net result of this is that the microrocket pump translates in the direction of the dense side of the microrocket pump at the velocity $v_r - v_m$. The huge pressure inside the core times the cross sectional area is a large force. Since the area separating the dense region from the semi-dense region translates, the force is a *thrust* which does work.

Particles fed into the dense side initially are translating at a range of velocities with a mean speed of v_m, same as they had in the background. Also, their RMS speed is v_r, same as they had in the background. However, when these particles get solidified as they are pressed into the microrocket pump without changing their energy their single velocity is v_r, the background RMS speed. Thus, their transport velocity changes from v_m to v_r, a 9% increase.

This solid assembly of brutinos must be of a special size and structure to provide stability. Also it propels the assembly slowly (at approximately 9% of the mean speed of the background particles).

The structure of the neutrino, i.e., the flow pattern, is complex. There are three distinct regions of the neutrino:

1. An almost spherically symmetric subsonic flow into a sink whose inner boundary consists of an almost spherical surface of radius r_c, the sonic sphere radius. The lack of symmetry results since the inhomogeneous state translates at 9% of the mean velocity of the background particles.

2. A compression chamber whose boundaries are defined by the sonic sphere at the inflow and the semi-solid portion of the *core* at its outflow.

3. The core consists of a very dense cylindrical region whose longitudinal axis passes through the center of the sonic sphere and which is parallel to the translational velocity vector of the assembly. The core radius is very small compared to the sonic sphere radius. Particles flow into the sides of the core. The core consists of semi-solid aft flowing particles in the trailing portion and a leading portion which at the sonic sphere center begins as semi-solid forward particles changing to a solid region. The particles exit the aft end of the assembly at velocity v_m and the forward end at velocity v_r.

4. The mass flow rate out the aft end at velocity v_m is greater than out the front at velocity v_r in order to balance linear momentum.

The whole assembly translates at a velocity slightly greater than c (the speed of light) where[2]

2   The factor 0.999720882 results form our guess that the velocity of light is slightly smaller than the speed of the neutrino. Incidentally this quantity 0.085378 is a value close to the electromagnetic coupling constant 0.085425 appearing in quantum electrodynamics.

$$c = 0.999720882\,(v_r - v_m) = 0.999720882\,(\sqrt{3\pi/8}\;v_m - v_m)$$

$$= 0.085378045 v_m$$

(3.12)

The semi-solid portion of the core acts like a microrocket with propellant and piston sections. This piston produces the suction for pulling particles into the sink. The individual particles entering the compression chamber have an average velocity of v_m and a flow velocity of 0.7 v_m. The spherical compression chamber can be considered as made up of a large number of triangular-cross-section cones. The particle velocity is directed toward the apex of each cone, i.e., the center of the sonic sphere, and their initial flow velocity is $0.700 v_m$. The particles entering have *zigzag* paths immediately after entering the compression chamber and some exit the sides of each cone and an equal number enter. As the flow progresses inward the transverse motion decreases until it is practically zero when it reaches the core. During this process the particle average velocity remains unchanged at v_m. As the particles approach the core the flow velocity approaches v_m. The flow volume from the subsonic flow region is proportional to the square of the sonic sphere radius. The sonic sphere radius is a direct function of the cross sectional area of the micro-rocket. The rocket size must be small enough so that the input to the compression chamber is *free molecular flow*. This means the sonic sphere radius is in the order of the mean free path. We describe each of these three sections in detail.

Unified Theory of Physics

E. OVERALL VIEW OF THE FLOWS

Let us look at the overall picture of the inhomogeneous assembly now. Figure 3.3 shows flow velocities inside the inhomogeneous assembly. The piston, shown as the rectangle, is like a *wave*. Particles are continually added to the aft end of the piston and it continually sheds particles from the front. As a result of this, adding to the aft end and shedding from the front, the piston particles move at v_r but the assembly travels at $v_r - v_m \approx c$.

The vectors in figure 3.3 show flow velocities for groups of particles. The assembly translates slowly compared to the inflow velocity at the sonic surface. Since the exiting forward flow velocity is v_r and the aft flow velocity is v_m the assembly travels at $v_r - v_m$. Since the assembly moves forward at velocity $v_r - v_m$ the pressure on the cross sectional area times the area is a force and that force moves at the velocity $v_r - v_m$. Thus, this force is a *thrust* and does work. It is the origin of all usable energy in this universe. The sonic surface will be almost spherical.

The analyses here, for the most part, assume the surface is spherical. With this assumption the inflow pattern is spherical and the mass flow rate, $\dot{m}_i$, is

$$\dot{m}_i = \rho_c A v = \rho_c 4\pi r_c^2 0.700 v_m \qquad (3.13)$$

where ρ_c is the mass density at the entrance to the sonic sphere.

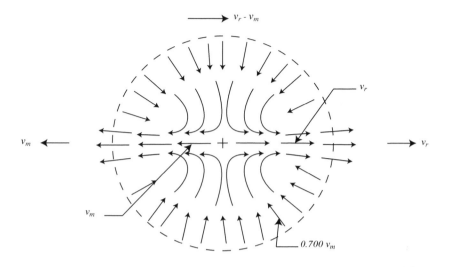

Figure 3.3. Inhomogeneous Assembly (Flow Vectors)

Once the configuration discussed here has been assembled we assume that it will be stable. The size of the assemblage, as measured by the sonic sphere radius, is determined by the cross section of the core. There exists a size of cross section which is stable. If the core cross section size increases, the inflow is interrupted with the result that the core sloughs off particles. If the core size is below the optimum size, then the core will grow. Possibly there are discrete sizes of the configuration. The size of the core is determined by the mean free path. The mean free path must be very large compared to the core diameter, otherwise the outflows from the front and the rear of the assemblage will disrupt the inflow and disassemble the configuration. We would not expect long time stable assemblages of air since ℓ/r for air is $10^{-7}/10^{-10} = 1000$. This is contrasted with an $\ell/r = 10^{-16}/10^{-34} = 10^{18}$ for this brutino gas.

This analysis ignores induced rotation about the transport velocity vector as the particles flow into the sink. This rotation results when the tangential pressure becomes smaller than the radial pressure. Such rotation will produce angular momentum. The superposition of rotary flow complicates the arguments here but it should not affect the stability of the assembly.

F. THE SUBSONIC FLOW REGION

A three-dimensional fluid sink causes a spherically symmetric flow toward the center of the sink. We can analyze sink flow by considering a sphere made up of a number of converging nozzles. The nozzles can have any cross section. The simplest cross section would be to use equilateral triangle cross sections. The inside surfaces of the nozzles would be frictionless since when the nozzles are all assembled to produce a sphere, the sides confining the flow are unnecessary. We present the analysis of the flow of a compressible fluid, i.e., gas, down a converging nozzle.

Figure 3.4 shows a converging nozzle draining a large reservoir of gas. The nozzle output goes into a receiver tank. The pressure p_r in the receiver tank can be controlled by a valve which dumps the contents into a vacuum space. This nozzle acts like a portion of a sphere with an indefinitely large radius. Thus, p_1, ρ_1, and v_{r_1} are assumed invariant. Also, the flow velocity v_1 is assumed to be zero irrespective of what occurs in the receiver tank. If p_r is equal to p_1 there is no flow. If p_r is reduced below p_1 then the gas will begin flowing into the receiver tank. The nozzle acts insulated and frictionless since the gas flow is spherically symmetric for sink flow.

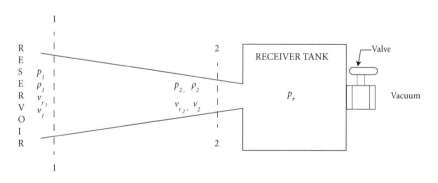

Figure 3.4. Concept of a Converging Nozzle
Emptying into a Receiver Tank

The analysis for this flow is presented on pages 298-300 of [3.1]. The flow velocity at the throat is

$$v_2 = \sqrt{\left(\frac{2k}{k-1}\right)\frac{p_1}{\rho_1}\left[1-\left(\frac{p_2}{p_1}\right)^{\frac{k-1}{k}}\right]} \tag{3.14}$$

where k is the ratio of the specific heat at constant pressure to that at constant volume. For our (ideal) gas $k=5/3$. Further $p=\frac{1}{3}\rho v_r^2$, where p is the pressure, ρ is the density, and v_r is the RMS velocity. Thus v_2 can be written as

$$v_2 = \sqrt{\frac{5}{3}v_{r_1}^2\left[1-\left(\frac{p_2}{p_1}\right)^{\frac{2}{5}}\right]} = 1.291\left[1-\left(\frac{p_2}{p_1}\right)^{\frac{2}{5}}\right]^{\frac{1}{2}}v_{r_1} \tag{3.15}$$

As the pressure is reduced it reaches a critical value p_c. Below this pressure the mass flow rate will stop increasing. The critical pressure is

$$p_c = p_1\left(\frac{2}{k+1}\right)^{\frac{k}{k-1}} = \left(\frac{3}{4}\right)^{\frac{5}{2}}p_1 = 0.487p_1 \tag{3.16}$$

The critical flow velocity is the local speed of sound. Its value is

$$v_c = \sqrt{\frac{kp_c}{\rho_o}} = 0.745\,v_{r_c} \tag{3.17}$$

Also

$$\frac{p_1}{\rho_1^k} = \frac{p_c}{\rho_c^k} \quad\text{and}\quad \rho_c = \left(\frac{p_c}{p_1}\right)^{\frac{1}{k}}\rho_1 = (0.487)^{\frac{1}{k}}\rho_1 = 0.649\rho_1 \tag{3.18}$$

Finally

$$p_c = \frac{1}{3}\rho_c v_{r_c}^2 \tag{3.19}$$

Now

$$v_{r_c} = \sqrt{\frac{3p_c}{\rho_c}} = \sqrt{\frac{3\times 0.487p_1}{0.649\rho_1}} = 1.50\sqrt{\frac{1}{3}v_{r_1}^{\,2}} = 0.866v_{r_1} \quad (3.20)$$

and

$$v_c = 0.745v_{r_c} = 0.745(0.866v_{r_1}) = 0.645v_{r_1}$$
$$= 0.645\sqrt{3\pi/8}\; v_{m_1} = 0.700v_{m_1} \quad (3.21)$$

Summarizing we have

$$v_c = 0.700v_{m_1}$$
$$p_c = 0.487p_1$$
$$\rho_c = 0.649\rho_1 \quad (3.22)$$
$$v_{r_c} = 0.866v_{r_1}$$

G. THE COMPRESSION CHAMBER

The neutrino is the result of a localized condensation of the gaseous *ether* which pervades the universe. The ether gas is made up of small elastic spherical particles (r_b = 4.052 × $10^{-35}m$), with a mass m_b = 2.89 × $10^{-66}kg$, and having an average velocity v_m = 3.51 × $10^9 m/s$. The ether particles make everything in the universe. The neutrinos make all the observables in the universe.

The condensation is produced by a microscopic ($10^{-25}m$ diameter) *pump* which is randomly built starting with a few ether particles. The ether particles have the same distribution of speeds as the background gas but are aligned to all move parallel to each other in the same direction. Thus, the group translates at velocity v_m. As a result of the *pump* other particles are attracted to the pump and the assemblage grows. The particles collide with these few particles, squeeze them together, and accelerate them to the velocity v_r without changing the energy of the group. This squeezing together is accomplished by increasing the speeds of some particles and decreasing the speeds of other particles.

As a simple example of the squeezing mechanics, consider the problem of two particles A and B with mass of unity each and with numerical velocities. Let the velocities be designated by the particle names. Let A have a velocity of 6 and B have a coincident linear velocity of 14 in the same sense, see Figure 4.2a.

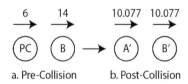

a. Pre-Collision   b. Post-Collision

Figure 3.2. Particles Squeezed Together at Same Energy

We want to get A and B to translate together at the same velocity and with the same energy A and B had when separate. We accomplish this by impacting A with C and B with D, where all four particles have colinear velocities.

The pre-impact mean velocity of A and B is $(6 + 14)/2 = 10$. The RMS velocity is $\sqrt{(6^2 + 14^2)/2} = 10.077$ The post impact velocities are primed and have the magnitude 10.077. To accomplish this we impact A with C and B with D. Figure 3.3 shows the impacts.

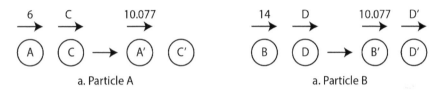

a. Particle A   a. Particle B

Figure 3.3. Impacts of A and B to Obtain Post-Impact Energies Same as the Particles Had Before Squeezing Together

For particle A

Momentum	$C+6=C' +10.077, \; C'=C-4.077$
Energy	$C^2+36=C'^2+101.55=C^2-8.154C+118.172$
	$8.154C=82.17, \; C=10.077, \qquad C'=C-4.077=6$

For particle B

Momentum	$14+D=10.077+D' \qquad D'=D+3.923$
	$D'^2=D^2+7.846D+15.390$
Energy	$196+D^2=101.55+D'^2=101.55+D^2+7.846D+15.390$
	$7.846=79.06, \; D=10.077, \qquad D'=14$

The four velocities are shown in Figure 4.4.

$$\overset{\longrightarrow}{10.077+} \quad \overset{\longrightarrow}{6} \qquad\qquad \overset{\longrightarrow}{6 \ + \ 10.077}$$

$$\textcircled{C} \quad \textcircled{A} \quad = \quad \textcircled{C'} \quad \textcircled{A'}$$

Figure 3.4. The Velocities For Making Two Particles Have the Same Velocity While Not Changing their Energy

Note that the velocities just interchanged. The momentum of A and B increased by 2(10.077)-20=1.54 while C and D decreased by 2(10.077)-20=0.154.

A neutrino can begin to be formed by a few ether particles getting randomly aligned to translate as a group at a velocity approximately equal to v_r. Such a phenomenon as this can occur simply from the random motion of the ether properties. Such groups, undoubtedly, are formed frequently and most are dissipated. However, occasionally, some assemblies survive and grow. Such an assembly, with its velocity being close to v_r, can start collecting particles by behaving like a *pump* and pulling a few particles toward it. The *pump* produces an inflow, and like any sink, the few incoming particles will tend to flow radially inward initially, but as they get close to the sink, they will begin circulating about the longitudinal centrally located velocity vector. The few assemblies that get this far without being disintegrated will begin increasing the central core which initially began as a few aligned particles and began to grow as the core grows. The entering particles begin to encounter less particles with opposing velocities. Eventually the particles come to the core with a mean velocity v_m, the value of the background mean velocity. Some are moving forward, some aft. The percentage of forward to aft is controlled by linear momentum conservation. The size of the *core*, which is assumed to have a cylindrical shape, begins at a very small diameter, possibly 10^{-33} *meters*. If the structure survives this far, the incoming particles will begin squeezing the forward stream.

This stream initially has a distribution of velocities. The particles are squeezed together so that their velocities are the same. As a result of this squeezing together, the forward moving particles all move at the same velocity and that velocity is v_r, if it is assumed that the squeezing process is a constant energy process. The aft end of the forward moving portion of the core will move at the velocity v_r-v_m. The pressure times the core area is a moving force, i.e., a thrust. This movement tends to produce a vacuum which *sucks-in* more particles and the core grows.

As the core grows, the mass flow, $\dot{m}$, into the core grows. At some stage in this growth, the neutrino becomes a viable structure. The mass inflow rate becomes large enough to produce a strong rotational flow, as measured by the magnitude of its angular momentum. This rotational flow aligns the particle velocities parallel to the assembly flow vector.

Let us consider the flow through the surface of a sphere whose center is at the center of the core. As the neutrino grows, the flow rate through this sphere reaches the local speed of sound. The mechanism of achieving sonic flow is that the particles entering this sphere have a mean free path possibly close to the sphere radius. As a result, a particle entering one side of the sphere will be unlikely to encounter a particle entering the opposite side. For the most part, the entering particles will encounter particles whose velocities are directed toward the sphere center. Recall that two steady streams of particles are continually exiting this sphere. These streams have such small cross sections that they do not interfere with the inflow. The streams are solid (the forward stream) or near solid (the aft stream) so that their cross-sectional areas are possibly ten orders of magnitude less than the *sonic* sphere radius.

As time passes, the core continues to grow and concomitantly the sonic sphere grows. After the sonic sphere radius becomes somewhat larger than the mean free path, the number of particles having mean free paths small compared to the sonic sphere radius will build a *back-pressure* which limits the inflow. This back pressure at the outlet of an air nozzle emptying into a vacuum limits the flow velocity. Thus, the core mass stops growing when it reaches a size on the order of $10^{-25}m$.

The sonic sphere size is determined by the magnitude of the mean free path. There are three recognized neutrino masses (electron, muon, and tauon). According to our theory the proton neutrino is a fourth mass. It is not known why there are these different mass nor what their detailed shapes are.

IV. THE STABLE FUNDAMENTAL MATTER PARTICLES

A. THE PROTON AND THE FINE STRUCTURE OF THE ELECTROSTATIC FIELD

The proton and the electron are the only two matter particles made of only one orbiting neutrino. They, and their antiparticles, are the only stable fundamental matter particles. The proton consists of a single neutrino. This neutrino has a mass equal to the proton's mass, 1.6726×10^{-27} kg. The proton is formed when a neutrino of this mass is impacted by other massive neutrinos and knocked into a circular orbit. The rocket-like thrust developed by the neutrino, which thrust is the same for all neutrino sizes, balances the neutrino centrifugal force. There is precisely one value of mass for which the neutrino thrust and centrifugal force will balance and have an angular momentum of $\hbar/2$ (the angular momentum the neutrino had before being knocked into orbit). That mass is the proton.

The proton angular momentum is

$$\hbar/2 = m_p r_p c \tag{4.1}$$

from which

$$r_p = \hbar/(2m_p c)$$

$$= \frac{1.0545716 \times 10^{-34}}{2 \times 1.672621 \times 10^{-27} \times 299792458} \tag{4.2}$$

$$= 1.051545 \times 10^{-16} \ m$$

The centrifugal force balancing the thrust gives

$$T=m_p c^2/r_p$$
$$=\frac{1.672621\times10^{-27}(299792458)^2}{1.051545\times10^{-16}} \quad (4.3)$$
$$=1.4295\times10^6 \text{ Newtons}$$

This is a large force!

The thrust is produced by the flow into the neutrino sink. The flow velocity at the entrance to the sonic sphere with radius r_c is $0.700v_m$ and the density is $0.649\rho_0$. Thus the mass flow rate $\dot{m}$ is

$$\dot{m}=(0.649\rho_0)(4\pi r_c^2)(0.700v_m)=5.709\rho_0 r_c^2 v_m \quad (4.4)$$

The mass flows into the neutrino center and the net momentum forward is $\dot{m}$ times the velocity jump from v_m to v_r. Thus the propelling force is

$$F= \dot{m}(v_r-v_m)=5.709\rho_0 r_c^2 v_m(\sqrt{3\pi/8}-1)\,v_m$$
$$=0.4876\rho_0 r_c^2(3.5103\times10^9)^2=6.008\times10^{18}\rho_0 r_c^2 \quad (4.5)$$

Equating this to the centrifugal force (4.3) and solving for $\rho_0 r_c^2$ gives

$$\rho_0 r_c^2=\frac{1.4295\times10^6}{6.008\times10^{18}}=2.379\times10^{-13}\ kg/m \quad (4.6)$$

Thus, we have a relation between the background density and the sonic sphere radius. Our analysis of the strong nuclear force gives us r_c. Thus, we are able to determine the background density, one of our four basic constants.

The proton is difficult to produce and is practically indestructible. It cannot decay by the usual ways of statistical variations — it must be destroyed by the method of its production, or by encountering an anti-proton. All matter particles consist of one, or more, orbiting neutrinos. An orbiting neutrino is a matter particle - an elementary matter particle.

The fundamental unit of matter is a single orbiting neutrino. We call this an *elementary matter particle*. Wherever there is matter, there is an orbiting neutrino. The neutrino is producing an almost spherical inflow of ether gas and sending out two fine streams of outflow, one at velocity v_r and the other at v_m. Additionally, anywhere there is an elementary matter particle there is a flow producing a measurable angular momentum of $\hbar/2$.

The proton *stands alone* as the largest elementary matter particle. A more massive orbiting neutrino would have an angular momentum greater than $\hbar/2$, which is impossible.

Many physicists believe the proton is made up of several other more basic particles. We think these other more basic particles they observe are either artifacts of their collision experiments and/ or are the *wake* flow produced by the single orbiting neutrino. We believe all matter particles made up of more than one elementary matter particle are unstable.

As the proton orbits, its neutrino produces an outflow of gas at velocity v_r followed by an opposite outflow of gas at velocity v_m. This results in waves with crest-to-crest dimensions in the order of $10^{-16}m$, the radius of the orbiting neutrino making the proton. The tangential motion in the field also produces $10^{-16}m$ sized waves. The resulting configuration consists of $10^{-16}m$ by $10^{-16}m$ by $10^{-16}m$ sized flows, which we call *wavespaces*. The wavespaces are the *fine structure* of the electrostatic field. The wavespaces move radially from the charged particle at the velocity of light c which is very slightly less than v_r-v_m. The wavespaces carry away the mass that is continually absorbed by the proton neutrino.

Figure 4.1 shows the fine structure of the proton field. Each of the wavespaces has a circulation flow which is right-handed for a positive charge. Right-handedness is determined by the counterclockwise flow of the neutrino making the proton.

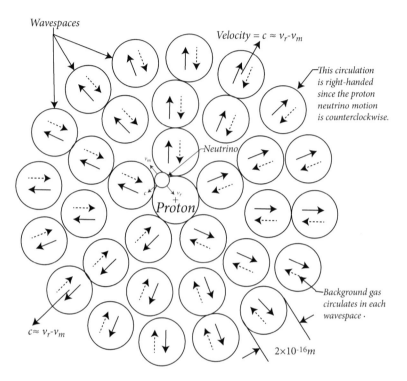

Figure 4.1. Fine Structure of the Proton Electrostatic Field.

B. The Strong and Weak Nuclear Forces

Let us now discuss the strong nuclear force between two protons (or nucleons). Each nucleon is assumed to consist of a proton sized orbiting neutrino. The orbiting neutrinos of the two nucleons must be of equal mass, which gives equal orbital radii, in order to develop the force. The weak nuclear force between two elementary matter particles can be produced if the strong nuclear force requirements are not met. The most likely cause is when the two elementary matter particles are side-by-side rotating in opposite directions.

Let us talk about the region between two proton parallel orbit planes through D and E separated by a distance *R* with a center C at *R/2* from each orbit plane (see Figure 4.2). There will be a general inflow toward the center of each proton. The basic ether particles, of course, are expelled from a very small area. The two protons will be *pulled* together (actually pushed together by the background pressure being higher than the pressure between the two because of their two *sinks*). They would take parallel orbits with their sinks as close together as possible. The limit on their closeness is established by the density increase as the background particles flow into the sinks.

Basically for the strong nuclear force we have two planes attracting each other and the sinks stay in synchronization as close to each other as possible—which stabilizes the orbits and makes them parallel. The spikes' exit flows are like four water hoses spewing out water as they rotate in parallel planes. These *rocket planes* are very thin and parallel, and are many spike cross-section diameters apart—possibly billions of diameters apart. We can estimate the force between these two such particles using an inviscid fluid. The analysis is further simplified, and less accurate, if we ignore density increase—which we will do here. As the nucleons get closer together, density effects begin to be significant and actually control the closeness of the nucleons.

We compute the attractive force between two hydrodynamic sinks by determining the flow velocity at the perpendicular plane through the bisection point between the two sinks, i.e., the plane A-A in Figure 4.2. The static pressure on this plane is the ambient pressure less the dynamic pressure i.e., $p_0 - (1/2)\rho v^2$. In this expression p_0 is the ambient pressure and v is the flow velocity. The reduction of pressure on this plane, compared with planes far removed from the sink, thus is taken as $(1/2)\rho_0 v^2$. The attractive force is the integral of this pressure over the plane A-A.

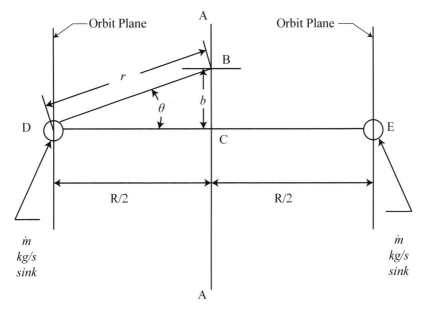

4.2 Hydrodynamic Flow Analysis of Two Sinks

The inflow of fluid is *ṁ* *kg/sec* (i.e., the sink strength). The mass inflow is $\rho A v$, where ρ is the density, A is the area, and v is the flow velocity. From mass continuity $\rho A v$ is constant and, for incompressible flow, $A v$ is constant. Let v_m be the mean background velocity. Let r_c be the value of radius at which the inflow speed [1] is $0.7v_m$. Now $r_c^2(0.7v_m)=r^2v$. Thus $v=0.7v_m\, r_c^2/r^2$.

The component of this velocity at the plane A-A parallel to the plane is directed from B to C in Figure 4.2, and its magnitude is $v\,sin\theta$. Due to the right sink, the component also is $v\,sin\theta$ so that the total flow is $2v\,sin\theta$. There is no flow normal to the plane due to symmetry of the sinks.

The pressure reduction now is

$$p=\frac{1}{2}\,p_0(0.7)^2\,v_m^{\,2}\,r_c^{\,4}/r^4 \tag{4.7}$$

The differential force is

$$pdA=\frac{1}{2}p_0(0.49)v_m^{\,2}r_c^{\,4}\frac{dA}{r^4} \tag{4.8}$$

The differential area is

$$dA=2\pi b\,db \tag{4.9}$$

Since $b= r\,sin\theta$ we have

$$db=r\,cos\theta d\theta$$
$$dA=2\pi\,r^2\,sin\theta cos\theta d\theta \tag{4.10}$$

Thus

$$pdA=\frac{1}{2}p_0(0.49)v_m^{\,2}r_c^{\,4}\,2\pi\,r^2\,sin\theta\,cos\theta\,d\theta/r^4 \tag{4.11}$$

---

1   This is the radius at which the inflow reaches (local) sonic speed, which is $0.7v_m$. We call this the critical speed, and the sphere it defines is called the sonic sphere whose radius is r_c.

Using

$$R/2=r\ cos\theta, \quad r=R/(2\ cos\theta) \tag{4.12}$$

the force then due to the sink flow of the doublet, which we label F_a, is

$$F_a=\int pdA=1.539\ \rho_0 v_m{}^2 r_c{}^4 \int_0^{\pi/2} \frac{4\sin\theta\cos^3\theta d\theta}{R^2}$$

$$=1.539\ \rho_0 v_m{}^2 r_c{}^4 \left[-\frac{4\cos^4\theta}{4R^2}\right]_0^{\pi/2} = 1.539\ \rho_0 v_m{}^2 r_c{}^4/R^2 \tag{4.13}$$

In front and behind the sink producing attraction, there are developed repulsive forces due to the spreading of the outflow of gas at the high velocity (v_r forward and v_m aft) which produce distant repulsive forces. The repulsive forces producing flows are modeled as spherically symmetric sources which are at a distance g from the (inflow) sink.

The net attractive force between the neutrino doublets now is the nuclear force and is given by

$$F_n=F_a-F_r=1.539\rho_0 v_m{}^2 r_c{}^4\left(\frac{1}{R^2}-\frac{1}{(R+g)^2}\right)$$

$$\approx 1.539\rho_0 v_m{}^2 r_c{}^4 \frac{2g}{R^3} \tag{4.14}$$

where we have assumed $R>>g$ to make the approximation. We note that except when the nucleons are very close that the strong nuclear force varies as the inverse cube of the separation distance.

Equation (4.14) implies that R can have any value just as R in the equation for the electrostatic force. However, two nucleons in equilibrium have a minimum value for R, i.e., their separation distance for which the force is zero. We denote this value by R_m. The experimental value for R_m is obtained from the general expression for the size of nuclei, which is

$$d=1.2\times10^{-15}A^{1/3}m \qquad (4.15)$$

In the case of ^1H the atomic weight A is 1 so that

$$d=1.2\times10^{-15}m \qquad (4.16)$$

We simply assume R_m is d.

This value of R_m is related to r_c, the sonic sphere radius. The two neutrinos making the two combined nucleons are pulled together by the strong nuclear force. However, as their separation distance reaches $2r_c$ their density begins to increase inversely with the squares of their half separation distance. Soon the increased density results in an increased pressure between the particles and eventually the attractive force will be zero. Further decreases in the separation will result in a repulsive force. The separation distance can be determined quantitatively. However, we have not done that. Instead we assume that when $R/2$ is $0.8r_c$ the force will be zero. Thus

$$r_c=\frac{R_m}{2(0.8)}=\frac{1.20}{1.6}\times10^{-15}=7.50\times10^{-16}m \qquad (4.17)$$

With this value of r_c we can now determine the background density from (4.6)

$$\rho_0=\left(\frac{2.379\times10^{-13}}{7.50\times10^{-16}}\right)^2=4.23\times10^{17}\ kg/m^3 \qquad (4.18)$$

Now we know that a neutrino in orbit carries around its sonic sphere of radius $7.50\times10^{-16}\ m$.

Let us now return to the nuclear force between two nucleons.

$$F_n = 1.539 \rho_0 v_m{}^2 r_c{}^4 \frac{2g}{R^3} \tag{4.19}$$

The binding energy of two nucleons, ^2H, is one megaelectron volt, or 1.602×10^{-13} *Joules.* Now

$$\int_{R_m}^{\infty} F_n dR = 1.539 \rho_0 v_m{}^2 r_c{}^4 \int_{R_m}^{\infty} \frac{2g}{R^3} \, dR = 1.539 \rho_0 v_m{}^2 r_c{}^4 \frac{g}{R_m{}^2}$$

$$= \frac{1.539 \times 4.23 \times 10^{17} \left(3.5103 \times 10^9\right)^2 \left(7.50 \times 10^{-16}\right)^4 g}{\left(1.20 \times 10^{-15}\right)^2} \tag{4.20}$$

$$= 1.76 \times 10^6 g = 1.602 \times 10^{-13}$$

Thus

$$g = \frac{1.602 \times 10^{-13}}{1.76 \times 10^6} = 9.10 \times 10^{-20} \ m \tag{4.21}$$

This is a very small distance but it is still much larger than the (square) core diameter. The strong nuclear force between two nucleons is

$$F_n = 1.539 \rho_0 v_m{}^2 r_c{}^4 \times 2 \times 9.10 \times 10^{-20}/R^3$$

$$= 2.80 \times 10^{-19} \rho_0 v_m{}^2 r_c{}^4/R^3 \ Newtons \tag{4.22}$$

Substituting for r_c, v_m, and ρ_0 we have

$$F_n = 2.80 \times 10^{-19} \times (3.5103 \times 10^9)^2 \times (7.50 \times 10^{-16})^4 \, 4.23 \times 10^{17}/R^3$$

$$= 4.62 \times 10^{-43}/R^3 \ Newtons \tag{4.23}$$

For R_m, (1.2×10^{-15}), the force is 267 Newtons.

We have not been able to determine the magnitude of the weak nuclear force.

C. THE ELECTROSTATIC FORCE

The flow into the (neutrino) sink and the v_r outflow from the front of the neutrino and the v_m flow from the aft end of the neutrino produce the strong nuclear force for two close protons. The flows at velocity v_r out the front followed by the flow from the aft end of the neutrino at a distance produce the (Coulomb) electrostatic field. We now discuss this field.

In this Newtonian universe of hard particles making up an ether, the assemblages of these particles (i.e., neutrinos) orbiting at the speed of light in circular orbits make up all matter at rest. The assemblages making up matter (i.e., the neutrinos) all have angular momentum which is either right-handed or left-handed. The different handedness makes the difference between positive and negative electrostatic charge. The orbiting assemblage making up a matter particle produces a pulsation in the ether like that of a breathing sphere. Two such assemblages whose centers are at a distance R apart can produce an inverse square force of interaction between them. The maximum magnitude of the force produced, from Bassett [4.1], is

$$F_e = \rho_0 \frac{8\pi^2 a^2 b^2 \alpha\beta}{T_e^2 R^2} \tag{4.24}$$

In this equation a is the nominal radius of one breathing sphere, b is the nominal radius of the other, α is the half amplitude of oscillation of one sphere which is taken as the proton orbital radius, β is the half amplitude of the other sphere, T_e is the period of charge oscillation, ρ_0 is the background mass density (which Bassett had taken as unity), and R is the separation distance. For more background see Whittaker [4.2]. With electrostatic charges all charges are alike except for the sign which, in this kinetic particle theory here, is controlled by the direction of rotation of the charge-producing assemblage (i.e., the neutrino) about its orbital tangential velocity. Thus, we set $b=a$ and $\beta=\alpha$ so that

$$F_e=\rho_0\frac{8\pi^2a^4\alpha^2}{T_e^2R^2} \qquad (4.25)$$

The half amplitude of oscillation α is taken equal to r_p. The period of oscillation is given by

$$T_e=2\pi r_p/c \qquad (4.26)$$

Now

$$F_e=\rho_0\frac{2a^4c^2}{R^2} \qquad (4.27)$$

The interaction with the background of this breathing sphere is characterized by the mean free path. As the sphere oscillates in and out it sends a pulse by its *average* interaction with the background, and that is by a sphere with a radius equal the mean free path. Thus, we take a to be the mean free path, ℓ. Equation (4.27) with ℓ replacing a gives the electrostatic force. The force also is e^2/R^2 so that

$$\rho_0\ell^4=\frac{e^2}{2c^2}=\frac{\left(1.5189\times10^{-14}\right)^2}{2\left(2.9979\times10^8\right)^2}=1.28349\times10^{-45}\ kg\text{-}m \qquad (4.28)$$

Using ρ_0 from (4.18) we solve for ℓ. Thus

$$\ell=\left(\frac{1.28349\times10^{-45}}{4.23\times10^{17}}\right)^{1/4}=2.35\times10^{-16}\ m \qquad (4.29)$$

We note that

$$\frac{\ell}{r_c}=\frac{2.35\times10^{-16}}{7.50\times10^{-16}}=0.313 \qquad (4.30)$$

As a check on the foregoing description of the mechanism of electrostatics, we developed the following model and analysis. To simulate electrostatic charge we made up a model of an arm rotating about a fixed axis and then having a dumbbell rotating on the end of the arm, as indicated in Figure 4.3. The charge in Figure 4.3 is positive since the rotation is right-handed. The arm length a simulates the orbital radius of the proton, and ω_1 is the angular velocity of the neutrino making the proton.

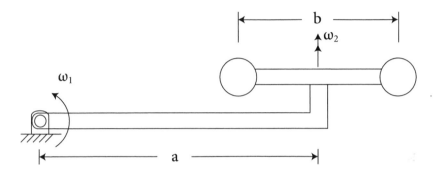

**Figure 4.3. Rotating Dumbbell Simulating
the Proton Charge Mechanism**

The *dumbbell* with arm of length b rotating at angular velocity ω_2 simulated the *spin* (i.e., rotary motion) of the neutrino making the proton. If ω_2 were in the opposite direction then it would simulate the opposite charge. We were able to prove that the rotation of the bar at angular velocity ω_1 with the simultaneous rotation of the dumbbell at angular velocity ω_2 relative to the main arm all immersed in an acoustic medium would produce a disturbance in the acoustic medium similar to that produced by a breathing sphere with a twist flow. A pair of these located a long distance r ($r >> a,b$) from each other would produce a force of repulsion between the two devices. If the rotation ω_2 in one device were reversed then the force would be attractive. From this analysis we concluded that an orbiting neutrino would produce the electrostatic field.

If a pair of rotating dumbbells of opposite polarity[2] were placed

---

2   Polarity is defined by ω_2 in Figure 4.3 being directed upward (positive) versus downward (negative).

so that they orbited each other and then another orbiting pair of opposite polarity were placed in a similar orbiting configuration then they would attract each other with a force varying inversely with the square of their separation distance. We concluded that this was the mechanism of gravitation. These results were derived by Dr. L. A. Steinert and published as an appendix to the Mississippi State University report on a US Army contract. See Brown [4.3].

D. THE ELECTRON

When the proton is formed it is necessary to form another structure to balance the effect on the background of the proton. The most obvious balancing structure is the one producing the negative electrostatic charge field. The structure also must have angular momentum of $\hbar/2$. Finally, it is necessary that the structure be such that it can have an existence separate from the proton. These properties, of course, are properties of the electron.

Our best model of a structure satisfying the above requirements for the electron is presented now.[3] Figure 4.4 shows the electron structure which is actually just the path taken by the extremely small mass (having a core cross section in the order of $10^{-51}\ m^2$) making the electron[4].

The dimensions on the drawing of Figure 4.4 are r_{em} having a value in the order of $10^{-19}\ m$, r_{es} in the order of $10^{-17}\ m$, and r_{ea} in the order of $10^{-11}\ m$. The smallest loop, with radius r_{em}, is the inertial balancing loop where the assemblage large thrust force is balanced by the centrifugal force of the electron mass. The next larger loop r_{es} is the loop r_{ea} balancing the electrostatic charge of the proton, and the large circular loop, with radius r_{ea}, is the loop producing the angular momentum of the electron (i.e., the spin of magnitude $\hbar/2$).

---

3   This model was first published on page 156 of Brown [4.4] and is shown on the front cover of that reference.

4   The cross sectional area of the electron (core) is estimated by squaring the proton core radius $(1.44\times10^{-25})\times4\pi\times$ the electron/proton mass ratio to the 2/3 power. Thus $A_e=4\pi(1.44\times10^{-25})^2(1/1836)^{2/3}=1.74\times10^{-51}\ m^2$. This assumes a *square* core.

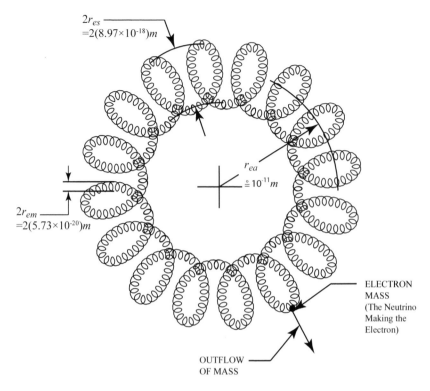

$2r_{es}$
$=2(8.97\times10^{-18})m$

r_{ea}
$\cong 10^{-11}m$

$2r_{em}$
$=2(5.73\times10^{-20})m$

ELECTRON
MASS
(The Neutrino
Making the
Electron)

OUTFLOW
OF MASS

Figure 4.4 Electron Structure

We present those analyses of the loops which we have been able to develop.

The electron orbital (or *mass*) radius r_{em} is controlled by balancing the assemblage thrust (having the same value as the proton assemblage thrust) with the centrifugal force produced by the electron mass m_e. Since the electron neutrino and the proton neutrino both translate at the speed of light the radii are proportional to their masses. Thus

$$r_{em} = r_p m_e / m_p = 1.051545 \times 10^{-16} / 1836.152668$$
$$= 5.7268951 \times 10^{-20} \ m \tag{4.35}$$

Simultaneous with the formation of the small orbital inertial balancing structure, a structure must be formed to balance the positive electrostatic field of the proton. The electrostatic field component is produced by a loop made up of the inertia balancing paths. This field component must be a wave with the same period as that produced by the proton. The propagation speed of this loop is smaller than the speed of light, obviously. We guess that its speed is $\sqrt{\alpha}\,c = (c/11.7)$. If its path radius is $1/11.7$ times the proton radius then the wave produced will have the same period as the proton's. Thus, the electrostatic wave produces the fine structure and is consistent with the proton electrostatic field.

The velocity of propagation of the electrostatic loop around the circle (the angular momentum circle) is much slower than the speed of light. This slower speed would necessitate that the angular momentum radius be greater than the radius required if the electron were moving at the speed of light. This results since the angular momentum is produced by the electron mass traveling in its nominal circular path of radius r_{ea}. We assume that the propagation velocity around the largest loop, or circle, is $\alpha c = 3\times10^8 / 137 = 2.18\times10^6$ *m/s*. Thus we compute r_{ea} from

$$\hbar/2 = m_e\, r_{ea}\, \alpha c \qquad (4.32)$$

or

$$r_{ea} = \left(\frac{\hbar}{2m_e \alpha c}\right) = 1.05\times \frac{10^{-34}\times137.1}{2\times9.11\times10^{-31}\times3\times10^8} \qquad (4.33)$$

$$= 2.63\times10^{-11}\ m \quad (=r_B/2)$$

where r_B is the Bohr radius.

This model of the electron is conjectural but is based on some fundamental facts. Any piece of matter must consist of at least one orbiting neutrino. Otherwise, there is no known way of controlling

49

the large thrust of the neutrino making matter. We are fairly certain that the electron consists of one orbiting neutrino (having the mass of an electron, of course). We also are fairly certain that the electron has a large orbit to produce its angular momentum. However, we are not certain of how the resonance is constructed which produces the electrostatic field.

We have not discovered what determines the mass of the electron. However, we think the electron may be made of the smallest neutrino possible. It could be that the electron neutrino is due to a resonance of the gas flowing into the neutrino. Incidentally the mass of the electron is given by

$$m_e = m_p/(6\pi^5) = m_p/1836.118109 \qquad (4.34)$$

However, no one has been able to derive this equation. The determination of the proton mass is straightforward. The elementary particles have an angular momentum of $\hbar/2$. The angular momentum requirement stemmed from the neutrino mechanism whose angular momentum is $\hbar/2$.

E. THE FIVE FORCES OF NATURE

The strongest force in nature is the thrust that propels the neutrino through the dense ether. Its value is 1.43 meganewtons. The second strongest force is the force holding nucleons together. Its value is 267 newtons, some four orders of magnitude less than the neutrino thrust. Next is the electromagnetic force. Its magnitude is 1/137.1 times the strong nuclear force and, thus, $1/10^6$ times the strength of the neutrino thrust. The force whose strength is next lower is the force involved in nuclear decay. It is called the weak nuclear force and it is six orders of magnitude less than the electromagnetic force, or 12 orders of magnitude smaller than the neutrino thrust. Finally, the fifth force is the force of gravitation, which is 38 orders of magnitude less than the electromagnetic force — or 44 orders of magnitude less than the neutrino thrust. The gravitational field is produced by the intersection of an electron orbiting a proton, as will be discussed in Chapter V.

The neutrino thrust drives a solid ball of mass with a diameter of 10^{-25} meters through a background gas whose density is ten trillion times that of lead. The neutrino is a *self-consistent* flow of background ether particles which consists of aligning background particles and squeezing them together so that they are all moving parallel to each other without changing their energy. The *squeezing together* operation produces the meganewton thrust and its velocity is $v_r - v_m$, where v_r is the RMS velocity of the background gas and v_m is the mean velocity. This thrust is the source of all usable energy in the universe.

The primary importance of the neutrino thrust is for its role in holding matter together. All matter in the universe is made of neutrinos orbiting in circular paths (when at rest) and in plane spiral paths when translating. The neutrino thrust balances its centrifugal force as it takes its orbital path. Neutrinos are made with varying masses and are right-handed or left-handed. The light mass neutrinos making matter will have small radii and massive ones will have large radii.

All orbiting neutrinos produce an *electrostatic field* as a result of their output, during orbit, of a flow at velocity v_r followed by a flow at v_m, which is 9% less than the flow at velocity v_r. The field produced is characterized by three-dimensional waves whose three dimensions are equal to the orbital radius of the orbiting neutrino producing the field. We call these three-dimensional spaces *wavespaces*. They travel at the velocity of light which is very slightly less than v_r-v_m and in radial directions from the emitting atom. (The emission in the two output streams at velocities v_r and v_m is due to the absorption of background gas by the neutrino.) All stable matter is strictly due to the proton. The proton makes the electron and, thus, the hydrogen atom. Nucleons are made of hydrogen atoms, and thus make all the larger atoms. The orbital radius of the proton is 10^{-16} m. Its wavespaces, thus, are 10^{-16} m by 10^{-16} m by 10^{-16} m and its electrostatic field is called the *Coulomb field*. The proton is the largest elementary matter particle.

The strong nuclear force is generated by the interaction of two equal mass neutrinos rotating *side-by-side* in the same direction and *in phase*. The most prevalent occurrence of the strong nuclear force is between the protons in nucleons. Every nucleon pair has two protons and the proton-proton binding holds them together. The binding is a result of the Bernoulli static pressure reduction as the background gas flows into the proton neutrinos. The increase in background density limits the closeness of the two nuclei.

The weak nuclear force is generated between two elementary matter particles. The neutrinos could be *out-of-phase,* of different size, of different rotational directions and still develop some force. It appears, from our analyses of the elementary matter particles (in this book), that the weak nuclear force occurring between elementary particles frequently is due to the particles rotating *side-by-side* in opposite directions.

The electromagnetic force is developed between two matter particles which are separated by a distance large enough for the matter particle field to act like a spherically symmetric breathing sphere. The Coulomb electromagnetic field is generated by the proton with its 10^{-16} *m* orbit. Polarity is due to the handedness of the orbiting neutrino.

The gravitational field is produced by two oppositely directed electrostatic fields. This is accomplished by the electrostatic fields of the electron orbiting about the electrostatic field of the proton with an amplitude equal to the radius of the brutino.

F. SYMMETRY IN PHYSICS

We now consider symmetries in physical theory. Let us discuss the *TCP* theorem. The three letters *TCP* stand for three hypothetical operations: *T*, time reversal; *P*, space reversal (which is approximately equivalent to taking a mirror image of space, and is called the parity operation); and *C*, charge conjugation, the technical name for interchanging particles and antiparticles.

We have just shown that charge conjugation and parity conservation are the same thing since all matter (and antimatter) are constructed only of neutrinos and antineutrinos. Parity is conserved since each elementary matter or antimatter particle is made of objects which are mirror images of each other. Thus, *C* and *P* are always conserved.

Time reversal just does not happen — even though it is theoretically possible. Time is simply a displacement. No displacement can be exactly reversed. Let us consider probably the simplest event imaginable. A brutino, *A*, is translating at exactly 10^9 meters per second, exactly parallel to the x-axis in Figure 4.5. Since time is simply a displacement, then for time to reverse, in this case, it is necessary that after collision the speed of *A* remain exactly at *v* and the direction be exactly reversed.

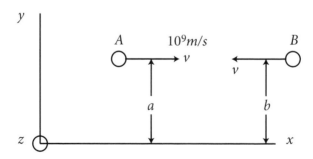

Figure 4.5. Brutino A Moving Parallel to x

We try to reverse A with a collision from particle B. Let us assume the speed of B, i.e., v, is exactly the same as the velocity v of A and the direction of B is exactly parallel to x. Further, assume the masses are exactly the same and both particles are perfectly elastic. (We've made a lot of impossible assumptions.) In order for A to return in its original path, distance b must be exactly equal to a. The probability of b being exactly equal to a is zero, even though we know it is theoretically possible.

A gross type of time reversal could occur when organic systems are completely understood. It might be possible to take an old monkey and reverse all the aging processes and return it to the sperm and egg. However, the detailed locations of the brutinos in a *super microscope* a microsecond after reversing the body would never be like it was a microsecond before starting the reversal. An additional thought: the monkey after one year of reversing would look like she did two years before.

Time reversal of a sort occurs, according to the kinetic particle theory of the universe. The universe began as a homogeneous Maxwell-Boltzmann gas. Condensations of the gas, caused by an entropy-reducing mechanism at the molecular size level, produced neutrinos. The neutrinos produced atoms with gravitational fields. Gravitation organized matter (made stars, earth, bacteria, plants, animals). Stars continually grew and finally got so large that they exploded, converting some matter back to neutrinos but most of the results of the explosions are matter of varying sizes from stars down to space dust and individual atoms. Thus, we see that the matter in the universe goes through a long cycle (possibly much greater than 10^{10} years). It is not clear what the cycle is for neutrinos. They presumably *live* for more than 10^{10} years. It is not clear whether or not the neutrino population goes through a cycle.

G. The Supposed Violation of Parity

With classical (Newtonian) theory, for anything which occurs, a mirror image can occur. This fact is known as the conservation of parity. Recently (1956), T. H. Lee and C. D. Yang claimed to have found theoretical evidence that parity is not always conserved and later that year (December, 1956) a group led by Ms. C. D. Wu and Ms. E. Ambler performed an experiment using Cobolt^{60} which confirmed the Lee and Yang prediction. The world physics community accepted that parity was not conserved. If this were true then the world is not a classical mechanics world as put forth in this book. In the following pages we will show the flaw in the argument that Cobalt^{60} decay violates the conservation of parity.

Cobalt^{60} is a radioisotope of cobalt which emits electrons. The decay is by the *weak* interaction. According to Lee and Yang's theoretical analysis, they predict that more electrons will be emitted opposite the spin vector than in the spin vector direction. Figure 4.6 shows a Cobalt^{60} atom, the nuclear spin vector (directed downward) and the current which consists of flowing electrons. The rotational vector for the current is upward, opposite the nuclear spin direction. Wu and Ambler performed such an experiment with Cobalt^{60} and determined that more decay electrons were flowing up than down. This experiment verified the Lee and Yang prediction. The emission of decay electrons is in the direction of the right hand thumb when the fingers are in the direction of the flowing currents.

Figure 4.7 shows a mirror image of the presumed current (flowing clockwise when viewed from the top, i.e., from the positive end of the z-axis). The spin axis shown in Figure 4.6 is not shown since it is not a physical entity and therefore would not appear in a mirror image. Since the presumed current vector is downward, the emitted electrons would go downward and not be a mirror image of Figure 4.6.

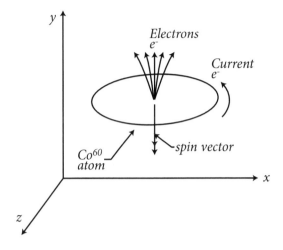

Figure 4.6. Cobalt[60] Experiment Showing Decay Electrons Emitted Opposite Nuclear Spin Direction

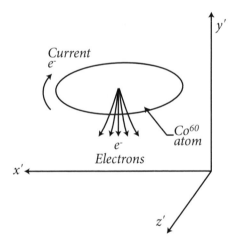

Figure 4.7. Presumed Mirror Image of Cobalt[60] Experiment Showing Mirror Image of Current Flowing Counterclockwise Looking Down

A mirror image of the Cobalt[60] experiment has never been performed, for obvious practical reasons. Actually, such an experiment has not been *thought out* or the physics community would realize the flaw in the interpretation of the actual Cobalt[60] experiments. First, and foremost, it is not conceivable that the mirror image of an electron is also an electron. The electron would have to be symmetric in order for its image to be an electron. It is hard to conceive of a particle model which is symmetric and has polarity. With the kinetic particle theory of physics, an antiparticle of a particle is simply the mirror image. All matter is made of neutrinos with handedness. Antiparticles are obtained from particles simply by changing the handedness. Of course, the mirror image of a neutrino changes the handedness. Thus, the mirror image of an electron is a positron.

With all this discussion we now show a theoretical experiment with a mirror image of a Cobalt[60] atom, which we believe is an Anti-Cobalt[60] atom. Figure 4.8 shows the current, which is the flow of positrons and has the direction of the mirror image of the current in Figure 4.6. The decay particles are the mirror image of the decay shown in Figure 4.6. They are positrons and are directed opposite the spin vector, which we show.

The flaw in the interpretation of the Cobalt[60] experiment is that the mirror image of an electron is not an electron but a positron. Thus, we conclude that the Cobalt[60] experiment did not violate parity.

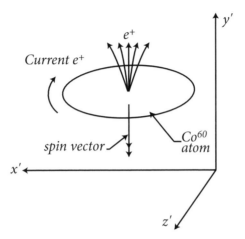

Figure 4.8. Actual Anti-Cobalt⁶⁰ Experiment Showing Decay Positrons Emitted Opposite Nuclear Spin Direction

V. The Hydrogen Atom, the Fine Structure Constant, Gravitation, and Cosmic Rays

A. The Hydrogen Atom and the Fine Structure Constant

The hydrogen atom is formed concomitantly with the proton and the electron. The proton is formed by the chance collision of a proton-massed neutrino by other massive neutrinos. As the proton neutrino goes into orbit, the electron is formed to balance the electrostatic field produced by the proton. The electron ends up orbiting the proton to produce the hydrogen atom.

We now compute the electron orbital velocity, for Hydrogen one (^{1}H). Balancing the centrifugal force with the electrostatic force gives

$$\frac{m_e v^2}{r_{eg}} = \frac{e^2}{r_{ep}^2} = \frac{e^2}{r_{eg}^2 \left(1 + m_e/m_p\right)^2} \tag{5.1}$$

where r_{eg} is the distance from the electron to the hydrogen center of gravity and r_{ep} is the distance from the electron to the proton

$$v^2 = \frac{e^2}{m_e r_{eg} \left(1 + m_e/m_p\right)^2} \tag{5.2}$$

60

The orbital angular momentum is[1]

$$\hbar = m_e r_{eg} v \tag{5.3}$$

So that

$$v = \frac{e^2}{m_e r_{eg} v \left(1 + m_e/m_p\right)^2} = \frac{e^2}{\hbar \left(1 + m_e/m_p\right)^2} \tag{5.4}$$

Neglecting the small term $(1 + m_e/m_p)^2$ and dividing by the speed of light gives.

$$\frac{v}{c} = \frac{e^2}{\hbar c} = \alpha \tag{5.5}$$

The term $e^2/(\hbar c)$ is known as the fine structure constant α and is the ratio of the electrostatic force to the strong nuclear force. The electrostatic force is proportional to the background density times the square of the electromagnetic force velocity, i.e., the speed of light c. The strong nuclear force is proportional to the density times the square of the nuclear force velocity, i.e., the background mean speed v_m. Thus we can conclude

$$\left(\frac{v_r - v_m}{v_m}\right)^2 = \frac{\text{electromagnetic force}}{\text{strong nuclear force}} = \left(\frac{\text{electromagnetic velocity}}{\text{nuclear force velocity}}\right)^2 \tag{5.6}$$

and this ratio is close to the fine structure constant. We know that the parameter $e^2/(\hbar c)$ is the fine structure constant α, its value is

---

1 The reason that the orbital angular momentum is $\hbar$ is because the velocity of the electron, before being captured by the atom is $2\pi\hbar/m\lambda$ (as will be shown later) and λ must be the circumference of the electron orbit. Larger, slower orbits can occur while satisfying the above equations. In addition, elliptical orbits can occur.

0.00729735253 (see [5.1]), and its reciprocal is given by (5.7), see Mohr [5.1].

$$\alpha = \frac{e^2}{\hbar c} = 1/137.0359998 \tag{5.7}$$

Further

$$\left[\left(v_r - v_m \right)/v_m \right]^2 = (v_r/v_m - 1)^2 = (\sqrt{3\pi/8} - 1)^2$$
$$= 1/137.108733 \tag{5.8}$$

Thus

$$\alpha \approx \left[\left(v_r - v_m \right)/v_m \right]^2 \tag{5.9}$$

Also we note that the parameter $(v_r/v_m - 1)^2$ predicts the orbital velocity of the electron in speed of light units, in its lowest energy state, and in ^{1}H to within 5 parts per ten thousand compared to the fine structure prediction. This results since the electron velocity actually is

$$\frac{v_e}{c} = \frac{e^2}{\hbar c} \frac{1}{\left[1 + m_e/m_p \right]^2} = \frac{0.007297352533}{\left[1 + 1/1836.15267 \right]^2}$$
$$= 0.0072894105 \tag{5.10}$$

since the proton is non-stationary. We also have

$$0.99972088(v_r - v_m) = c$$
$$v_m = \frac{c}{0.99972088 \times \sqrt{3\pi/8} - 1} = \frac{299792458}{0.99972088 \times 0.085401882}$$
$$= 3.511353066 \times 10^9 \tag{5.11}$$

$$\left[\left(v_r - v_m \right)/v_m \right]^2 = \left(\sqrt{3\pi/8} - 1 \right)^2 = 0.007293481$$

Finally

$$\frac{\left[\left(v_r - v_m\right)/v_m\right]^2 - v_e/c}{\left[\left(v_r - v_m\right)/v_m\right]^2} = \frac{0.007293481 - 0.007289410}{0.007293481}$$

(5.12)

$$\frac{0.000004071}{0.007293481} = 0.00055817 \text{ or 5 parts in 10,000}$$

The discovery of this relationship was reported by Brown, Harmon, and Wood [5.2]. The significance of this discovery was that it showed the relationship of physical observations to the postulates of this kinetic particle theory of physics. It lends credence to our claim that the universe consists only of kinetic particles.

One of the most significant electrostatic field interactions is the electrostatic attraction between the electron and the proton in the simplest hydrogen atom. In the formation of the hydrogen atom, it is not known why the electron is formed with the mass it has and, as a consequence, why the orbital radius has the magnitude of the Bohr radius.

B. GRAVITATION

We assume that gravitation is produced by a similar *breathing sphere* mechanism as that producing electrostatic charge. When the proton is formed the electron is formed to balance the flows produced by the proton. Thus the negative charge field of the electron balances the positive charge field of the proton — insofar as possible. The absolute limit of this balancing is controlled by the diameter of the brutinos making up these two fields. We anticipate that this imbalance would result in a spherical breathing mode with a single amplitude of r_b, the radius of the brutino. Even though we do not have a clear understanding of the model we can make a guess at the mechanism which is consistent with the values for the parameters of the system.

First of all we note that a residual field produced by a pair of two opposite polarity fields should produce a field without polarity. The gravitational field has no polarity.

We begin with the equation for the interaction force F_g of two identical breathing spheres.

$$F_g = \frac{2\rho_o a^4 \alpha^2 v^2}{r^2 R^2} \tag{5.13}$$

where ρ_o is the background density, a is the nominal radius of the sphere, α is the half amplitude of the oscillation, v is the maximum velocity of the oscillation, r is the radius generating the oscillation period ($T=2\pi r/v$), and R is the separation distance of the two atoms. The period is defined by the proton radius. Thus $r=r_p=1.05\times10^{-16}m$. The half amplitude is r_b, the brutino radius, from the assumption that the two fields are balanced within the limit imposed by the *graininess* of the fields. The velocity v clearly is c. Thus, our only unknown is a. Let us solve for a.

$$F_g = G\frac{m_H^{\,2}}{R^2} = \frac{2\rho_o a^4 r_b^2 c^2}{r_p^2 R^2} \tag{5.14}$$

where m_H is the mass of a hydrogen atom and G is the universal gravitational constant, $G= 6.673\times10^{-11}$ m^3/s^2-kg. From this

$$a = \left[\frac{Gm_H^{\,2}r_p^2}{2\rho_o r_b^2 c^2}\right]^{1/4} = \left[\frac{6.673\times10^{-11}\times\left(1.673\times10^{-27}\right)^2\left(1.052\times10^{-16}\right)^2}{2\times4.23\times10^{17}\left(4.05\times10^{-35}\right)^2\left(2.998\times10^8\right)^2}\right]^{1/4}$$

$$= 3.589\times10^{-16}\ m \tag{5.15}$$

This value is over fifty percent greater than the mean free path

$$\frac{a}{\ell} = \frac{3.589\times10^{-16}}{2.35\times10^{-16}} = 1.53 \tag{5.16}$$

This value of a given by (5.15) probably means that something is slightly incorrect in the gravitational model. The diameter of the effective sphere, according to our analysis, is 1.53 times as large as the breathing sphere diameter in the electrostatic analysis. It is not clear what the problem is. To illustrate the discrepancy let us compute G using $a=l=2.35\times10^{-16}m$ instead of $3.598\times10^{-16}m$.

$$F=2\times4.23\times10^{17}(2.35\times10^{-16})^4(4.05\times10^{-35})^2$$
$$\times(2.998\times10^8)^2/[(1.05\times10^{-16})^2R^2] \tag{5.17}$$
$$=G(1.673\times10^{-27})^2/R^2$$

Thus

$$G=1.23\times10^{-11}m^3/(kg^2\text{-}s^2) \tag{5.18}$$

The measured value of *G* is

$$G=6.673\times10^{-11}m^3/(kg^2\text{-}s^2) \qquad (5.19)$$

which is five times larger than our estimate. There are uncertainties in our result due to the simplicity of our modeling of gravitation. Much refined modeling can be accomplished which we hope would further substantiate our model of gravitation.

Equation (5.14) is the equation for the force of gravity which results from the postulates of the kinetic particle theory of physics. Equation (5.14) is an equation similar to (4.25) but it gives the electrostatic force which results from the same set of postulates that produced the gravitational force equation. Thus, the kinetic particle theory of physics should give the *unified field theory* sought by Einstein and others.

C. Origin of Cosmic Rays[2]

Cosmic rays are matter and radiation particles which are very energetic, are numerous, and shower the earth continually. Large quantities of the particles have energies in the range of $10^8 ev$ (or 10^{-11} *Joules*). However, extremely infrequently the earth will experience a particle with an energy of $10^{20} ev$. It is known that their origin is nearby, certainly within our visible region of the universe. This results from the fact that there is enough matter (space dust) in space to break down the rays. The physics community *has no idea* what the origin is of these particles.

We have a good idea from where the particles come, at least the low energy ones. There are almost a million neutrinos per cubic meter of space – at all times. A few of these neutrinos have masses close to the mass of a proton. When one such neutrino collides, and begins orbiting, it produces a hydrogen or an anti-hydrogen atom – the chance of either is equally-likely. In the human cosmos, almost all orbiting neutrinos are matter, rather than anti-matter. Thus, when an anti-hydrogen atom is made, because of its gravitational field, it soon is attracted to a matter atom and is disintegrated. The energy of its disintegration products is mc^2 ($10^{-26} \times 10^{17} = 10^{-9}$ *Joules*, or 10^{10} *ev*). We do not know how to calculate the frequency of such events, but we know that the production of hydrogen and anti-hydrogen is the source of all matter in the universe.

The origin of the high energy ($10^{20} ev$) cosmic rays possibly is due to a large number (10^{10}) of anti-hydrogen atoms escaping disintegration by being formed in a region of our space where gravity is very small. When they meet matter, they disintegrate and produce high energy cosmic rays.

---

2   This section is based on information presented in Chapter 11 of *We Have No Idea*, Cham and Watson, [5.3].

VI. RELATIVITY, THE WAVE PROPERTY OF MATTER, MAGNETISM, AND COMPARISON OF NEWTON AND EINSTEIN MECHANICS

A. ELECTRODYNAMICS AND RELATIVITY

In this book we have assumed a universe of identical elastic spherical particles which make up a gaseous ether and make up all matter and radiation. Further, we assume the gaseous ether is in a three dimensional space with a separate absolute time. The particles are very small with a radius in the order of the Planck length $\doteq 10^{-35}$ m and with a mass billions of billions times smaller than the electron. The particle number density is very large $\doteq 10^{83}$ m^{-3} and the mean free path is on the order of nuclear particle diameters.

A photon is assumed to be a stable assemblage of a large number of these ether particles translating at the speed of light, of course. Each fundamental matter particle at rest, such as a proton, is assumed to be a neutrino which is a very small stable assemblage (again made up of very many basic particles) moving at the speed of light in a circular path with a very small diameter.

From these above assumptions we immediately have the result that the energy of a matter particle at rest is the matter particle mass M_0 (which is the background particle mass times the number of particles making up the matter particle) times the square of the speed of light. Defining energy as mass times the square of the velocity of the mass we have

$$E_0 = M_0 c^2 \qquad (6.1)$$

which is the famous Einstein formula for the *equivalence* of mass and energy. Einstein assumed that mass could change to energy and vice versa. In the kinetic particle theory energy is mass in motion - nothing more, nothing less.

In order to accelerate matter, a series of photons bombard the matter, and each photon is partly scattered and partly captured. The captured parts of the photons are the mass that is added to the matter as it accelerates. When a force does work on a particle of mass *m* and accelerates *m* from zero velocity to *v*, the work done is $1/2mv^2$ so that $1/2mv^2$ is the particle's *kinetic energy*. The average energy of the accelerated mass during this process is the kinetic energy[1]. If a mass moving at velocity *v* (having energy mv^2) is absorbed by another mass moving at the same magnitude and direction of velocity *v* then the energy change of the increased mass particle is mv^2. Consider now a photon of energy $e_y = m_y c^2$. Consider the acceleration of a matter particle due to the result of scattering photons. Let M_0 be the mass of the matter particle at rest and let M_v be the mass when moving at velocity *v*. The matter particle energy when moving is $M_v c^2$ and the linear momentum is $P_m = M_v v$. The linear momentum P_y of the photons (assuming many photons were used) is $P_y = M_y c$, where M_y is the sum of the photon *masses*. Let the scattered photon total mass be $M_s = kM_y$ so that the captured mass is $M_c = (1-k)M_y$. The momentum transferred by the captured mass of each photon is $m_y(1-k)c$, where m_y is the mass of one photon. The average momentum transferred by the scattered portion of each photon is $m_y kc$ since the scattering is spherically symmetric with the maximum back scatter momentum being $2m_y kc$ and the minimum forward scatter being zero for an average of $m_y kc$. Thus, the momentum transferred to the matter particle is all of each photon's momentum. The total momentum

---

1   Appendix B presents a discussion of energy and kinetic energy.

imparted then is the sum of the initial momentum of each photon. Let us denote the total momentum as P. The differential energy change for the matter particle is the force times the distance so

$$dE_v = Fdx \qquad (6.2)$$

where E_v is the energy of the moving matter particle and F is the force applied. The force is the time rate of momentum change of the matter particle which, also, is the time rate of momentum change of the group of impacting photons. Thus

$$F = dP/dt \qquad (6.3)$$

Now

$$dE_v = Fdx = (dP/dt)dx = vdP \qquad (6.4)$$

We also have

$$dP = d(M_v v) = M_v dv + vdM_v \qquad (6.5)$$

and

$$dE_v = vdP = v(M_v dv + vdM_v) \qquad (6.6)$$

further

$$dE_v = d(M_v c^2) = c^2 dM_v \qquad (6.7)$$

Thus

$$c^2 dM_v = v(M_v dv + vdM_v) \qquad (6.8)$$

Simplifying

$$(dM_v)/M_v = vdv/(c^2-v^2) \qquad (6.9)$$

Integrating M_v from M_o to M_v and v from o to v gives

$$\ln\frac{M_v}{M_o} = -\frac{1}{2}\ln\frac{c^2-v^2}{c^2} = -\frac{1}{2}\ln(1-\beta^2) \qquad (6.10)$$

where $\beta=v/c$. Now

$$\frac{M_v}{M_o} = \frac{1}{\sqrt{1-\beta^2}} \qquad (6.11)$$

We see that this is the well-known mass growth equation and note it has been derived from classical Newtonian mechanics which uses an absolute space with a separate absolute time system.[2]

The fact that the velocity of matter can never exceed the speed of light results simply from the fact that the accelerating agent (i.e. the photon) is moving at the speed of light.

When a photon interacts with matter at rest the circular path becomes a planar spiral path as seen from a rest frame. However, in a frame moving at the translational velocity v of the particle the planar spiral is seen as a closed path and since angular momentum is constant

---

2   This analysis was developed for this theory by the late Dr. Darrel B. Harmon, Jr.

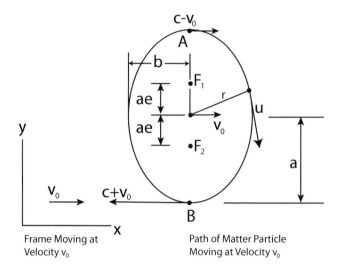

Figure 6.1 Matter Particle Moving at Velocity v_0

along the path, the closed path is an ellipse. Figure 6.1 shows the elliptic path of a matter particle moving to the right at velocity v_0. The two foci F_1 and F_2 are shown. The mass m takes the path shown by the ellipse in the reference frame moving at velocity v_0.

Since the mass always moves at velocity c (the speed of light) the mass at point A moves at velocity $c-v_0$ in this frame and the mass at B moves at velocity $c+v_0$. Angular momentum conservation requires that the mass at A times $(c-v_0)$ times the distance A to F_2 be the same as the same mass at B times $(c+v_0)$ times the distance F_2 to B. The major semi-axis is a, the minor semi-axis b, and the eccentricity is e, as shown in Figure 8.1. Angular momentum conservation gives

$$(a+ae)(c-v_0)=(a-ae)(c+v_0) \tag{6.12}$$

Solving for e gives

$$e=v_0/c=\beta \tag{6.13}$$

Thus, the eccentricity is the particle translational velocity in speed of light units.

We will now determine the relation between the orbit shape and the particle velocity. Figure 6.2 shows an elliptic orbit of a particle moving to the right at velocity v_0 . An ellipse is the locus of points where the distance from the point of a fixed focus (i.e. distance $\overline{AF_1}$) added to the distance from that same point to the other focus (i.e. distance $\overline{AF_2}$) is constant. For example, if a string of length $\overline{F_1A}$ plus $\overline{AF_2}$ is fixed at F_1 and F_2 and a pencil is placed inside the string then the trace will be an ellipse.

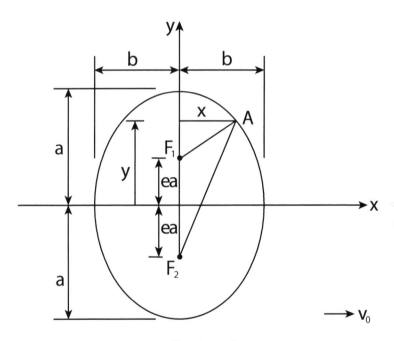

Figure 6.2 Elliptic Orbit Geometry

The length of the string is given by

$$\overline{F_2A} + \overline{AF_1} = \sqrt{(ea+y)^2 + x^2} + \sqrt{(y-ea)^2 + x^2} = 2a \quad (8.14)$$

To simplify this, let

$$z^2 = x^2 + y^2 + e^2 a^2 \tag{6.15}$$

Using z, eq.(6.14) becomes

$$\sqrt{z^2 + 2eay} + \sqrt{z^2 - 2eay} = 2a \tag{6.16}$$

Squaring both sides gives

$$z^2 + 2eay + \sqrt{z^4 - 4e^2 a^2 y^2} = 2a^2 \tag{6.17}$$

Moving the first two terms of (6.12) to the right side and squaring again gives

$$(2a^2 - z^2 - 2eay)^2 = z^4 - 4e^2 a^2 y^2 \tag{6.18}$$

Equation (6.18) simplifies to

$$(a^2 - a^2 e^2)(a^2 - y^2) = a^2 x^2 \tag{6.19}$$

or

$$(1 - e^2)(a^2 - y^2) = x^2 \tag{6.20}$$

When

$$y = 0 \quad x = \pm b \tag{6.21}$$

then

$$a^2 (1 - e^2) = b^2 \tag{6.22}$$

Thus

$$b/a = \sqrt{1-e^2} \qquad (6.23)$$

Since $e = v_o/c = \beta$, from (6.13), we have

$$b/a = \sqrt{1-\beta^2} \qquad (6.24)$$

This is the ratio of the minor axis to the major axis and clearly shows the orbit size reduction. Every matter particle in a piece of matter, such as a bar of steel, experiences this shortening with velocity. Thus, the complete bar will be shortened in the direction of motion by the factor $\sqrt{1-\beta^2}$. We therefore have

$$\ell_v/\ell_o = \sqrt{1-\beta^2} \qquad (6.25)$$

The velocity u of mass m on this elliptic path at radius r from F_2, as shown in Figure 6.1, is given in McClusky [6.1] by the equation

$$u^2 = g[2/r - 1/a] \qquad (6.26)$$

where g is a constant ($=GM$ by McClusky). The maximum velocity occurs when $r = a - ea = (1-e)a$ and has the value $c + v_o$. From this

$$(c + v_o)^2 = g\left[\frac{2}{(1-e)a} - \frac{1}{a}\right] = \frac{g}{a}\frac{1+e}{1-e} \qquad (6.27)$$

and

$$g = (c + v_o)^2\, a(1-e)/(1+e) \qquad (6.28)$$

Let the time for an orbit, i.e. the period, be τ_v (from reference [6.1]), substituting the value of GM as g from the above, and using e as β gives

$$\tau_v = 2\pi a^{3/2}/\sqrt{g} = 2\pi a/(c\sqrt{1-\beta^2}) \qquad (6.29)$$

When $v_o = 0$ (i.e. when $\beta = 0$) the period $= 2\pi r/c$ – obviously the circumference of the circle divided by the speed of light. The period increases with motion and grows without bound as $\beta(=v_o/c)$ approaches unity, or as the velocity approaches the speed of light. Nuclear particles, which disintegrate and emit radiation and produce other matter particles, are observed to decay slower when moving – and governed by the law, $\tau_v/\tau_0 = 1/\sqrt{1-\beta^2}$ where τ_v is the decay time while moving at velocity v and τ_0 is the decay time while at rest. If it is presumed that decay takes an average number of orbits at rest and that the decay process depends upon the number of orbits (i.e. the number of trials at *breaking loose*) then it follows that

$$\tau_v/\tau_0 = 1/\sqrt{1-\beta^2} \qquad (6.30)$$

This gives the time dilation effect produced by the special theory of relativity, but here derived from a classical Newtonian basis.

Let us illustrate the kinematics of the acceleration of a proton. Figure 6.3a shows a photon approaching a proton. The proton, of course, is made up of a neutrino orbiting at the speed of light and with a mass equal the proton mass. Figure 6.3b shows the subsequent path of the proton neutrino after impact as seen by an observer at rest relative to the pre-impact proton. Figure 6.3c shows the proton as seen by the observer moving at the post-impact velocity. This moving observer sees its path as an ellipse.

We note that the proton orbit horizontal width has reduced to $2r\sqrt{1-(v/c)^2}$ and that the time to orbit is increased as given by the equation $T_v = T_0/\sqrt{1-(v/c)^2}$.

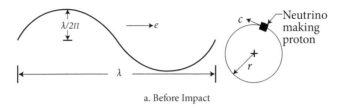

a. Before Impact

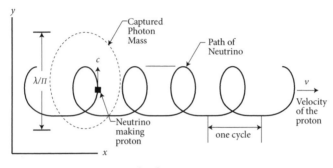

b. After Impact

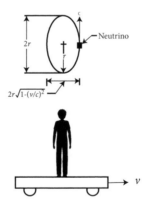

b. After Impact Collision by Observer Moving at Velocity *v*

Figure 6.3. Acceleration of a Proton by a Photon

B. MATTER WAVES

When any matter particle translates from one place to another along a nominal straight path, it always undulates from one side to the other as it moves. All matter at rest consists of elementary matter particles which consist of mass moving at the speed of light in circular paths. Ordinary matter, such as a bar of steel, is accelerated from rest by photons being transferred, usually from other matter (such as during impact by another bar). In the previous paragraphs we discussed how these photons interact with the flow fields produced by matter to accelerate the matter particles. When a small (low energy-long wavelength) photon interacts with a matter particle the distance from the particle center of mass to the coupling position must be in the order of the wavelength of the photon. Angular momentum considerations require that small impacts transfer mass at large distances from the center of the matter particles. Thus, the smaller the interacting photon energy (and the longer the wavelength) and the smaller the resulting matter particle velocity, the greater the eccentricity of the coupled mass. Now consider a free wheel in space rotating with a small unbalanced mass placed at a large radius. The axis of the wheel will undulate as it translates to keep the center of mass following an exact straight line. As a result the center of the wheel will take a sinusoidal path to the left and right of the center of mass straight path. Let us now calculate the wavelength of the moving matter geometric center path.

Let r_c be the distance from the matter particle's center to the place where the momentum is captured, by both the captured mass and scattered mass. The linear momentum P times the capture radius is Pr_c - which is equal to $\hbar$. For low velocities (i.e. non-relativistic conditions) the momentum P is also $M_0 v$, where M_0 is the matter particle rest mass. Now, we can write

$$\hbar=Pr_c=M_0\, vr_c\,, \qquad h=2\pi\hbar=M_0 v2\pi r_c=M_0 v\lambda \qquad (6.31)$$

where λ is the wavelength. Thus

$$\lambda=h/(M_0 v)=h/P \qquad (6.32)$$

This is the relation postulated by deBroglie and is called the *deBroglie wavelength*. For high speeds (i.e. for relativistic speeds) consideration must be given to matter particle mass growth, the center of gravity difference, and the matter particle contribution to the angular momentum.

The amplitude of the wave executed by the particle is very much smaller than the wavelength, in the order of 10^{-6} times the wavelength for some low energy impacts. We discuss this later in the section analyzing the translation of free electrons.

The wavelength λ is measured in meters, the constant h is 6×10^{-34} kilogram-meter^2 per second (i.e. Planck's constant), M_0 is the matter particle rest mass in kilograms, and v is the particle translational velocity v meters per second. An electron with a mass of 10^{-30} *kg* and a velocity 1/3 the speed of light (i.e., 10^8 *m/s*) has a wavelength of

$$\lambda=\frac{6\times10^{-34}}{10^{-30}10^8} =6\times10^{-12} \text{ meters} \qquad (6.33)$$

– a very small wavelength. The amplitude of the oscillation is smaller than the wave length by the factor of $1/2\pi$.

The observation of high speed mass growth with velocity is a significant part of high speed (relativity) physics and the observation of matter moving as a wave is a significant part of small item (quantum) physics. Both of these mechanisms come about simply from the interaction of a photon with matter – as shown by the mass growth formula and the wavelength formula just derived.

Throughout the 20th century many authors have stated the impossibility of deriving the special relativity results from classical (Newtonian) theory. We have shown that the three primary relativity observations (mass growth, matter shortening, and time dilation) are derived in a straightforward manner from a classical kinetic particle theory. Further, the mass-energy equivalence $E=M_0c^2$ is an obvious result of this theory. Finally, we have derived the wave properties of matter rather than postulating them – as done in contemporary physical theory. In summary, these results indicate that the universe is a classical based system.

C. MAGNETISM

Let us now determine the effect of motion on the electromagnetic force between two charged particles. An electron at rest, in the assumed kinetic particle universe, has an assemblage of kinetic particles orbiting at the speed of light in circular paths. In order to accelerate an electron, a photon with angular momentum $\hbar$ impacts the electron electrostatic field. The angular momentum of the combined assemblage (consisting of the electron and the captured portion of the photon) increases by $\hbar$ and the two entities combined translate at velocity v. The angular momentum then of the combined entities is $\hbar = mr_c v$, where m is the mass (of the two entities) and r_c is the half-amplitude of the center of the *charge*. The center of mass, of course, continues on a straight path. The undulation of the electron is the *electron wave*.

Consider now two like electric charges moving at velocity v parallel to each other and with a vector R starting at one charge and ending at the other and which vector is perpendicular to the particle velocities. In a reference frame moving at velocity v the two charges are seen to oscillate along the vector v. Assuming phasing is controlled by the twist component of the orbiting assemblage producing the charged particle, the maximum force of interaction between the two particles is given by the same formula as the formula for electrostatic charge. The difference is that a will be the deBroglie wave amplitude of the charge (which is $\lambda/(2\pi)$), and the period T_m will be $2\pi a/v = \lambda/v$. The force then due to motion will be

$$F_m = \rho_0 \frac{8\pi^2 a^4 \alpha^2}{T_e^2 R^2} = \rho_0 \frac{8\pi^2 a^4 \left(\lambda/(2\pi)\right)^2}{\left(\lambda/v\right)^2 R^2} = \rho_0 \frac{2a^4 v^2}{R^2} \qquad (6.34)$$

Dividing the magnetic force by the electrostatic force gives

$$\frac{F_m}{F_e} = \frac{2\rho_0 a^4 v^2 / R^2}{2\rho_0 a^4 c^2 / R^2} = \left(\frac{v}{c}\right)^2 \qquad (6.35)$$

for the special case of equal charges, equal and parallel velocities, and a charge separation vector initiating on one charge and ending on the other where the vector is perpendicular to both velocities. This ratio, of course, is the ratio of the magnetic force to the electrostatic force for this special case. Thus this mechanism models the magnetic force.

If the charges have the same sign then the force is attractive; if opposite, the force is repulsive. By the same mechanism, if the charge velocities are opposite, the repulsion/attraction is reversed due to the helicity of the charged particles.

Let us now generalize the special case just developed. Consider a velocity v_1, of charge 1 which produces 100 oscillations in a given period of time. Starting with velocity v_2, of charge 2 with 100 oscillations, if the velocity is reduced say to 1 oscillation in the same time period then the force, clearly, would be reduced to $1/100^{th}$ of the initial value. Thus, we can generalize the magnetic force equation to

$$F_m = (e^2/R^2)(v_1/c)(v_2/c) \qquad (6.36)$$

where v_1 and v_2 can be any value, negative or positive.

The next generalization is for unlike charge magnitudes. We let N_1 be the number of elementary charges at one location and N_2 at the other and take

$$q_1 = N_1 e \quad \text{and} \quad q_2 = N_2 e \qquad (6.37)$$

then

$$F_m = \frac{q_1 q_2}{R^2} \left(\frac{v_1}{c}\right)\left(\frac{v_2}{c}\right) \qquad (6.38)$$

If v_1 and v_2 are perpendicular to each other then the force would be zero because of the phasing and should vary sinusoidally from zero when perpendicular to a maximum magnitude when parallel (or anti-parallel).

Finally, if the radius vector R for the general case starts at charge 1 and ends at charge 2 (no matter what the relative locations and directions that the charges have) we have the magnetic force given by

$$F_{\sim m} = \frac{q_1 q_2}{c^2} \, v_{\sim 2} \times v_{\sim 1} \times \frac{i_R}{R^2} = \frac{q_1 q_2}{c^2 R^2} \times v_{\sim 2} \times v_{\sim 1} \times i_{\sim R} \qquad (6.39)$$

In this expression q_1 and q_2 are the point charges with units of $kg^{1/2}$ $m^{3/2} \, s^{-1}$, v_1 and v_2 are the charge velocities in m/s, i_R is a unit vector from charge 1 to charge 2, R is the magnitude in meters of the vector from charge 1 to charge 2, c is the speed of light in m/s, and $F_{\sim m}$ is the magnetic force in Newtons, which is attractive in the case where the velocities are parallel and equal and the charges are of like sign.

The above expression can be put in a more familiar form using the concept of the magnetic field. Let the magnetic field generated by charge 1 be

$$B_{\sim} = \frac{q_1}{c^2 R^2} \, v_{\sim 1} \times i_{\sim R} \qquad (6.40)$$

Now the magnetic force on charge 2 is

$$F_{\sim m} = q_2 v_{\sim 2} \times B_{\sim} \qquad (6.41)$$

The electromagnetic units also can be changed to Coulombs and Teslas, if desired.

Let us consider now the effect of relativity. All assemblages making matter, when at rest, orbit in circular paths. To accelerate a particle from rest, mass is added and the path is changed to a planar coil. If the planar coil is viewed from a frame moving with the center of mass of the moving particle then the path is elliptic. The time for an orbit is increased as given by the relation

$$\tau_v = \tau_0 / \left(\sqrt{1 - (v/c)^2} \right)$$

(6.42)

where τ_v is the period while moving and τ_0 is the period while at rest.

For two charged particles of like charge moving parallel to each other at absolute velocity v and where the vector connecting the two particles is perpendicular to v then the electromagnetic force between them is a force of repulsion given by

$$F_{em} = \frac{q_1 q_2}{R^2} \left[1 - \left(\frac{v}{c} \right)^2 \right]$$

(6.43)

If these charges are viewed from a frame moving at the same velocity as the charges then the separating force must be the same as given above. However, when seen in this moving frame the charged particle response would appear to be that due to a force

$$F'_{em} = \frac{q_1 q_2}{R^2}$$

(6.44)

since the particle velocities in this frame would be zero.

The particle response is experienced only by the acceleration and in this moving frame it would be $d^2y/d\tau_v^2$, if the y-axis is taken to pass through the two particles. The response then as measured by a clock at rest would be

$$\frac{d^2 y}{d\tau_v^2} = \frac{d^2 y}{d\tau_0^2} \left(\sqrt{1 - (v/c)^2} \right) = \frac{d^2 y}{d\tau_0^2} [1 - (v/c)^2]$$

(6.45)

Thus, the force would have to be reduced by the factor $[1 - (v/c)^2]$. If the charges are of opposite sign then the electrostatic force is attractive but the magnetic force is repulsive so that the same factor $[1 - (v/c)^2]$ results.

Einstein did not understand how the observed acceleration between two parallel moving electrons could depend upon the observer's velocity. This motivated him to develop the special theory of relativity. The difference in the acceleration measured was due only to the clock running slower when moving.

D. COMPARISON OF NEWTON AND EINSTEIN MECHANICS

I. INTRODUCTION

Around 1900 experiments on the velocity of light indicated that matter shrunk with velocity by the factor $\sqrt{1-\beta^2}$, where $\beta=v/c$, v is the velocity of matter, and c is the speed of light. Also, radioisotopes decayed slower when moving by the factor $1/\sqrt{1-\beta^2}$. This implied that distance measuring devices, such as a meter stick would shrink with velocity and clocks would run slower when moving.

As an example consider a meter stick moving at the velocity $0.866c$. For this, $\sqrt{1-\beta^2} = \sqrt{1-(0.866)^2} = 0.5$. Figure 6.4 shows a conceptual view of a meter stick with (conceptual) atoms.

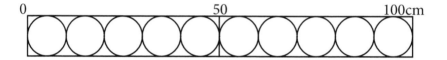

a. Meter Stick at Rest

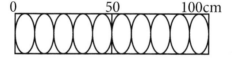

b. Meter Stick Moving at 0.866c

Figure 6.4. Rest and Moving Meter Sticks

As you can see the meter stick shrunk by the factor 0.5. Also the circular atoms squeezed down to ellipses. It is not obvious that the orbit times increased by the factor $1/\sqrt{1-\beta^2}=2$. We address this question later.

It turns out that the absolute velocity of light can be computed using instruments on moving frames as well as on rest frames. Let us consider a photon traveling for one second and a reference system traveling at 0.866c with respect to a rest frame. Figure 6.5 shows the geometry.

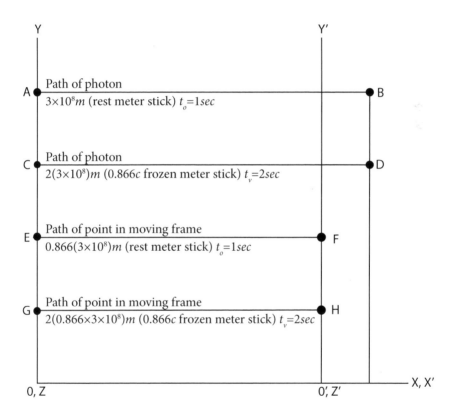

Figure 6.5. Displacements and Time Measurements with Rest and Moving Instruments

The figure shows a fixed coordinate system OXYZ and a moving reference frame O'X'Y'Z'. The moving frame started at 0 and translated to the right at velocity 0.866c for 1 second. The photons started at the same time. AB shows the path of the photon traveling $3 \times 10^8 m$ in one second. The distance $3 \times 10^8 m$ was measured with a non-moving meter stick and the time was measured using a resting clock. CD shows a photon traveling the same distance, but the distance and time are measured with meter sticks which were *frozen* while they were moving at $0.866c$, then fixed to the OXYZ system and laid down for $3 \times 10^8 m$ along the CD path. The length measured $2(3 \times 10^8)m$ and the moving clock measured 2 seconds. The path EF is for a point fixed to the O′X′Y′Z′ frame and distance was measured with fixed meter sticks and time was measured with a clock fixed to the OXYZ frame. The path GH was made by a point attached to the moving frame with $0.866c$ frozen meter sticks attached to the rest frame and the clock was fixed to the moving frame.

As you can see, the absolute velocity of the photon can be measured with resting or moving instruments. Also, the absolute velocity of the moving frame is $0.866c$, and can be determined with rest, or moving, instruments.

Since the absolute velocity of light can be computed by moving instrument readings taken on any reference frame, no matter what its velocity is, requires that matter behave in a certain manner. To determine the behavior we write expressions for the absolute velocity of light using measurements from two different reference frames, then equate the expressions, since they both represent the absolute velocity of light. This analysis is reported on pages 542-544 of Leigh Page [1]. The result of this analysis is that matter must shrink by the factor $\sqrt{1 - \beta^2}$ and orbit times must increase by the factor $1/\sqrt{1 - \beta^2}$.

In this analysis Page, like Einstein, assumed the light velocities were velocities relative to the moving frames - while they both were actually using absolute velocities. As a result Page and Einstein erroneously concluded that the matter shortening and time dilation were non-Newtonian phenomena. What this analysis actually shows is that matter shrinks and orbital periods increase but does not show that space shrinks nor time dilates with the motion of matter.

From these results we see that matter must shrink and orbit periods must increase when matter moves. Let us return to Figure 6.5. The coordinates of a <u>point</u> on matter (moving in the positive X and X` direction) for the OXYZ coordinate system are related to the coordinates in the O'X'Y'Z' system by the relations

$$x'=x\sqrt{1-\beta^2}$$
$$y'=y$$
$$z'=z$$
$$t'=t/\sqrt{1-\beta^2}$$

(6.46)

Einstein apparently noted that meter sticks shrink and orbit periods increase on moving matter in such a way that the absolute velocity of light could be computed accurately with the measurements from any moving frame - no matter what its velocity. However, he assumed that the velocity of light calculated from measurements on a particular reference frame was the velocity of light with respect to that frame. Thus, he thought the velocity of light relative to any moving frame was always $3 \times 10^8 m/s$. Thus he thought space and time were interconnected in a non-Newtonian manner. He concluded that space as well as matter shrink with velocity and universal cosmic time as well as orbit time, dilates with motion.

In conclusion, we note that matter in motion obeys Newtonian mechanics. However, Einstein's theory, applied to matter, gives the same results as Newton's theory. In applying Einstein's theory to space shrinkage/cosmic time dilation no measurements have ever been made to verify or deny the Einstein theory. Einstein's space-time theory is not true. Matter shrinks and orbital periods increase with velocity, but space does not shrink nor does time dilate with velocity.

II. Magnetic Forces and Mass Growth with Velocity

Figure 6.6 shows two electrons located at a distance r separating them. Figure 6.6a

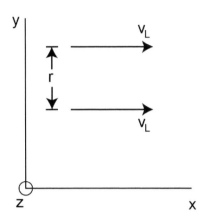

a. Electron Viewed in Laboratory

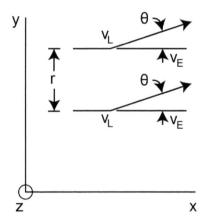

b. Absolute Velocity of Electron
Figure 6.6. Two Interacting Electrons

shows the electrons as viewed in the laboratory, where v_L is the velocity viewed. Figure6.6b shows the absolute velocities, $v_L+v_E\cos0$, where 0 is the angle between the two velocitires (which may not be in the plane of the paper) and v_E is the absolute velocity of the earth. It is noted that we do not know the magnitude or direction of v_E.

According to the Newton system the force between the electrons is a force of repulsion

$$F_N = \frac{e^2}{r^2}\left[1-\left(\frac{v_L+v_E\cos\theta}{c}\right)^2\right]$$ (6.47)

where e is the charge of the electron and e is the speed of light. For small values of $v_L(<<c)$ and assuming an earth velocity of $0.01c$

$$F_N' = \frac{e^2}{r^2}\left[1-(0.01)^2\right] = \frac{e^2}{r^2}(0.9999)$$ (6.48)

The force F_N' may not be experimentally distinguishable from F_N. For large values of $v_L(\approx1.0c)$ F_N' and F_N could be distinguishable except that the velocity can not be determined accurately since the velocity is determined by the measurement of the mass at velocity divided by the rest mass.

According to Einstein system the force is

$$F_S = \frac{e^2}{r^2}\left[1-\left(\frac{v_L}{c}\right)^2\right]$$ (6.49)

For velocities close to the speed of light, say $v/c=0.9$, the magnetic force is significant. Using

$$v = v_E + v_L\cos\theta$$ (6.50)

when v_L, for example in $0.99c$, would imply that v could be greater than c. However, determining v is done by measuring m_v/m_0 and solving for v using the mass growth formula

$$m_v / m_0 = \frac{1}{\sqrt{1-\beta^2}} \tag{6.51}$$

The experiment measures m_v/m_0 by applying electromagnetic forces to the electron beam then measuring its deflection on a screen. From this, subject to measuring errors of m_0, the value of v/c is the absolute speed of the electron.

Let us determine some values of m_v/m_0 for a range of assumed earth velocities.

v (meters/s)	$\beta = \frac{v_E}{c}$	β^2	$\sqrt{1-\beta^2} = \frac{m_0}{m_v}$	Mass Reduction	Experimental Detection
3×10^5	10^{-3}	10^{-6}	0.9999995	5 in 10^7	No
3×10^6	10^{-2}	10^{-4}	0.99995	5 in 10^5	Possibly
3×10^7	10^{-1}	10^{-2}	0.995	5 in 10^3	Yes

TABLE 6.1. Mass Variation with Earth Velocities

We know the velocity of the earth is at least 3×10^5 *m/s*. If the velocity of the earth were 3×10^5 *m/s* our experiments could not detect the reduction of mass. If the earth velocity were 3×10^6 *m/s* we possibly could not detect the reduction of mass. If the earth velocity were 3×10^7 *m/s* we probably could detect the mass reduction. These calculations show that in mass growth experiments whether v is taken as the absolute velocity of the electron or the laboratory value give results whose differences are undetectable. As far as mass growth with velocity is concerned Newtonian and Einsteinian systems are experimentally indistinguishable — just as with magnetism.

Let us consider the analysis of mass of matter growth with velocity. An electron is accelerated by impacting it with photons. When a photon impacts an electron its mass is partly captured and partly scattered—as the Compton scattering experiments show.

Also, just applying Newton's conservation laws of mass, linear momentum, and energy give the same results as the Compton scattering experiments, for each impact. The Newtonian acceleration analysis is presented on page 99-101 gives eq. (6.51).

As a prelude to our matter acceleration analysis, let us recall the Compton scattering experiments. In these experiments, a high energy photon impacted a charged particle. Part of the energy of the particle was captured and a lower energy photon was scattered off. The Newtonian advocate considers the photon as a one wavelength harmonic string of ether particles. Upon impact the charged particle captures part of the mass in a closed ring around the charged particle. The ring has an angular momentum of h, Planck's constant. The remaining mass re-forms into a longer one harmonic wavelength assembly, which is then scattered.

Incidentally, the average scatter angle is 90°, which is helpful in the mass growth analysis. Also, impacts of photons produce scattered mass which always moves at the same velocity as before the impact, namely c. Although this is not what Compton was searching for, the linear momentum of the captured mass was directly added to the linear momentum of the charged particle and the linear momentum of the scattered photon added momentum as a result of the change of direction. Thus, the charged particle was accelerated and had its mass increased. Repeated photon impacts would continue adding mass and continue increasing the velocity, always keeping the angular momentum at h.

It is difficult to conceive of a photon structure which can split and re-form into two photons, other than a string of elastic particles spread evenly along a one harmonic wavelength. Also the energy of the added mass is always mc^2, implying that the mass is stored in a closed ring around the nucleus. The angular momentum of the mass is always h. When new mass is added it combines with the old mass to make a new ring with its angular momentum remaining at h.

Similarly when some mass is lost its wavelength is increased so that the leaving mass has an angular momentum of h and the mass left behind increases its radius so that the angular momentum is h.

It is difficult to conceive of a carrier for the string of elastic particles making the photon other than waves in a gas. This, of course, implies that there exists a gaseous ether pervading the universe. Next, we ask where do the particles come from? The obvious answer is from the ether itself.

The resulting mass growth relation was first noted from high velocity accelerations of electrons. Einstein postulated eq. (6.51). Here again the Newtonian and Einsteinian systems are experimentally indistinguishable. Incidentally, it is obvious that a matter particle can not reach the speed of light since it is accelerated by photons which translate at the speed of light. It also is obvious that the mass of matter must grow without bound as its velocity approaches the speed of light.

III. THEORETICAL EXPERIMENTS TO DETERMINE EARTH'S VELOCITY

The formula for mass growth with velocity can be written as

$$\frac{m_v}{m_0} = \frac{1}{\sqrt{1 - \left(v_E + v_L \cos\theta\right)^2 / c^2}} \qquad (6.52)$$

where m_0 is the mass of the matter particle at absolute rest, θ is the angle between v_E and v_L, where v_E is the earth absolute velocity and v_L is the velocity of the particle relative to the earth.

Using eq. (6.52) there is a theoretical way to determine the earth's absolute velocity. Assume a value of v_E, for example, as $0.01c$ and use a value of v_L, for example $0.1c$, then perform experiments with directions varying over 4π steradians. There will be an angle maximizing m_v/m_0 and an angle minimizing m_v/m_0 differing from the maximizing angle by 180°. The angle making $v_E + v_L \cos\theta$ minimum will make m_v/m_0 a minimum. That angle is 180° from the velocity of the earth.

Using this angle in experiments we have

$$\frac{m_v}{m_0} = \frac{1}{\sqrt{1 - \left(v_E + v_L\right)^2 / c^2}} \qquad (6.53)$$

Now, by varying v_L there again will be a value of v_L that maximizes and another value which minimizes m_v/m_0. The values of v_L minimizing m_v/m_0 is the magnitude of the Earth's velocity.

The value then of m_0 can be determined from the expression

$$\frac{m_{earth}}{m_0} = \frac{1}{\sqrt{1 - \left(v_E/c\right)^2}} \qquad (6.54)$$

where m_{earth} is the mass of the particle determined on the earth.

We call this a theoretical method since such experiments, as required, would be difficult to make. Furthermore, the attainable accuracy probably would not be adequate.

IV. KINEMATICS OF THE MOTION OF MATTER

Einstein postulated that the energy of matter, at rest, is m_0c^2. Further, we know that

$$m_v - m_0 = \Delta m \qquad (6.55)$$

and that the energy increase is

$$\Delta E = \Delta mc^2 \qquad (6.56)$$

Thus, it seems safe to assume that

$$E_0 = m_0c^2 \qquad (6.57)$$

That is what Einstein did and that is what a Newtonian advocate would do.

There is only one way in a Newtonian world for matter at rest to have an energy of m_0c^2 and that must be for the mass m_0 to orbit in a circular path at the velocity of light. So, eq. (6.57) results from mass moving in a circle at the speed of light. We assume that all matter at rest is made up of elementary units of mass moving in a circular pattern at the speed of light.

Research made on the velocity of light, in different directions, is satisfied by matter shrinking with velocity as given by

$$l_v = l_0\sqrt{1-\beta^2} \qquad (6.58)$$

where l_v is the length of a bar of metal (for example) when moving at velocity v parallel to the bar axis, and l_0 is its length at rest.

Nuclear decay experiments show that the time for decay increases with velocity of the nuclear matter as given by

$$t_v = \frac{t_0}{\sqrt{1-\beta^2}}$$

(6.59)

where t_0 is the decay time at rest and t_v is the decay time when moving.

Let us now consider how matter, made up of a mass circulating at the speed of light, can be forced to translate. Since the velocity of the circulating mass must always remain at the speed of light, a method of motion is just to change the direction of the circulating mass.

Recalling the Compton scattering experiments, when a photon impacts a matter particle it adds mass to the particle. The matter particle must move. Since the mass making up the matter particle before impact was already moving at the speed of light, the only way for the particle to move is for the orbiting mass to change its direction. The mass changes from its circular path to a plane spiral path. Incidentally, the kinematics of motion of matter also limits the velocity of matter to less than the speed of light.

Let us now consider the paths made by the orbiting mass— making up matter particles. Figure 6.7 shows the paths for different situations. Figure 6.7a shows the circular path for a matter particle at rest. Figure 6.7b shows the path of a matter particle moving to the right at velocity v. Figure 6.7c shows the path for a moving particle when viewed from a frame moving with the particle.

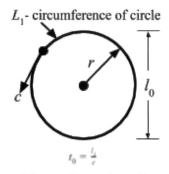

L_1- circumference of circle

r

c

l_0

$t_0 = \frac{l}{c}$

a. Matter Particle at Rest

l_3

a

b

c

$t_v = t_0 / \sqrt{1 - \beta^2}$

b. Matter Particle Moving at Velocity v

$$l_v = l_0 \sqrt{1 - \beta^2}$$

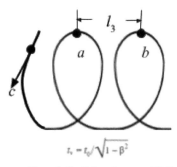

l_v

c

l_0

$t_v = t_0 / \sqrt{1 - \beta^2}$

c. Matter Particle Moving at Velocity v
as Seen By Observer Moving at v

Figure 6.7. Matter Particle in 3 Situations

In the reference frame moving with the particle (at absolute velocity v) the minor axis of the ellipse is parallel to its velocity and its length at velocity, l_v, is given by eg. 6.58. Also, the time to orbit, at velocity v, is given by 6.59.

An analysis of the paths follows: the velocity of the particle is given by

$$v = \frac{i_2}{t_v} \tag{6.60}$$

where t_v is the time it takes to travel the distance l_3. We can now write the velocity as

$$v = \frac{l_3}{l_v} = \frac{l_3}{t_0}\sqrt{1-\beta^2} = \frac{l_3}{l_1}c\sqrt{1-\beta^2} = \frac{l_2\beta c}{l_1}\sqrt{1-\beta^2} = \frac{v\sqrt{1-\beta^2}}{\sqrt{1-\beta^2}} = v \tag{6.61}$$

since

$$\frac{l_2}{l_1} = \frac{1}{\sqrt{1-\beta^2}} \tag{6.62}$$

This analysis shows the details of the path taken by the orbiting mass making a matter particle. The orbital analysis was fashioned after the planetary orbital analysis presented by McClusky[6.1].

This kinematic analysis shows the Newtonian basis for matter shortening and orbital time increasing as the velocity of matter increases. This, of course, agrees with the empirical relations available at Einstein's time.

V. SPEED OF LIGHT RELATIVE TO A MOVING FRAME

In this section we show how to determine the speed of light relative to a moving frame using clocks and meter sticks attached to the moving frames.

Figure 6.8 shows a fixed frame S and a frame S' moving at velocity $v_{S'}$ relative to the fixed frame. The velocity of the photon relative to the absolute frame in figure 6.8 is

$$v_{\gamma_s} = 3 \times 10^8 \ m/s$$

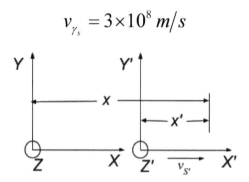

Figure 3. Photon Motion Relative to Two Frames

The photon velocity relative to the moving frame, by using the Newtonian (Galilean) system is

$$v_{\gamma_{s'}} = c - v \tag{6.63}$$

For example if v were $0.866c$ then

$$v_{\gamma_{s'}} = c - 0.866c = 0.134c = 0.134\left(3 \times 10^8\right) m/s \tag{6.64}$$

Let us now calculate the speed of light relative to a moving frame using measurements made with a meter stick on the frame and a clock moving with the frame. Let us do this for a frame moving at absolute velocity of $0.866c$. See figure 4. Let us freeze the meter stick while it is moving. Its length then will be half as long as a rest meter stick—now attach it to the rest frame. Start a photon at the zero of the meter stick and let it translate parallel to the meter stick. Put down frozen meter sticks to continue measuring the displacement of the photon. When the photon has traveled the measured distance of $2(3 \times 10^8)$ m the moving clock will measure 2 seconds. Figure 2 shows the geometry.

One observer measures the distance the photon traveled with a meter stick at rest and a clock at rest. Another observer B measures the distance the photon traveled with his frozen meter stick as $2(3 \times 10^8)$ m and the distance the X Y Z frame moved as $2(3 \times 10^8 \times 0.866)$ m. The upper numbers on the lines are those made with a rest meter stick and below with a frozen $0.866c$ meter stick.

Observer B now computes the velocity of the photon relative to him as

$$v = \left[2\left(3 \times 10^8\right) - 2\left(3 \times 10^8 \times 0.866\right) \right] / 2 = 0.134\left(3 \times 10^8\right) m/s \quad (6.65)$$

which is the same as given by (6.64).

If we measure a photon which travels an absolute distance of 3×10^8 meters with a frozen moving meter stick and a moving clock on a frame which is moving at velocity v then the photon absolute velocity computed will be

$$v = \left[\frac{3 \times 10^8}{\sqrt{1 - \beta^2}} \right] \div \left[\frac{1}{\sqrt{1 - \beta^2}} \right] = 3 \times 10^8 \, m/s \quad (6.66)$$

We note that this computation is independent of the velocity of the moving frame. Erroneously Einstein, and many others, interpret this result to mean that the speed of light is always 3×10^8 m/s relative to any frame.

VI. EINSTEIN'S THEORY

Einstein thought, that since using displacements and times recorded by instruments on a moving frame, that the computation of the speed of light with the frame instrument readings gave the speed of light relative to the frame. Thus, the photon relative velocity was independent of the frame velocity. This led him to his *principle of relativity*. His principle is *the laws of physics* are the same when determined relative to one inertial system as when determined relative to any other. For example, no matter how fast a space ship is going, an electron accelerator on the space ship would give a mass growth as given by eq. (7) where $\beta = v/c$ is the acceleration velocity, i.e., the velocity of the electron relative to the ship. If the ship were going at $0.6c$ and the electron were going at $0.9c$ relative to the ship, β in eq. (7) would be 0.9.

With this principle of relativity Einstein went on to derive the formulas (6.58) and (6.59). This result justified his assumption, as far as he, and the world-wide physics community were concerned.

Einstein's derivation of matter shortening and time dilation might have been the analysis presented on pages 543-4 of Page[6.2], which we present here.

Assume that the absolute speed of light has the same constant value in systems S and S', then the equation of a spherical wavefront diverging from the origin in system S is

$$x^2 + y^2 + z^2 - c^2 t^2 = 0 \qquad (6.67)$$

The equation of the same wave-front referred to system S' is

$$x'^2 + y'^2 + z'^2 - c^2 t'^2 = 0 \qquad (6.68)$$

From symmetry $y'=y$ and $z'=z$. Also x' and t' must be linear functions of x and t if the transformation is to be true for all locations in space and for all epochs of time. Therefore, we let

$$x' = k(x - at) \tag{6.69}$$

$$t' = l(t - bx) \tag{6.70}$$

where k, l, a, b are constants. Consider a point fixed in S' so that $dx'=0$. Then

$$dx = a dt, \quad a - \frac{dx}{dt} = 0 \tag{6.71}$$

The first of these linear relations, eq. (6.69), can be replaced by

$$x' = k(x - vt) \tag{6.72}$$

Now, substituting into eq. (6.68)

$$k^2\left(x^2 - 2vxt + v^2t^2\right) + y^2 + z^2 - c^2t^2\left(t^2 - 2bxt + b^2x^2\right) = 0 \tag{6.73}$$

or

$$x^2\left(k^2 - c^2l^2b^2\right) - 2x\left(l^2v - c^2l^2b\right) - c^2t^2\left(l^2 - k^2\frac{v^2}{c^2}\right) + y^2 + z^2 = 0 \tag{6.74}$$

This equation (6.74) must be identical with (6.67). Comparing the equations we see

$$k^2 - c^2l^2b^2 = 1, \quad k^2v - c^2l^2b = 0, \quad l^2k^2\frac{v^2}{c^2} = 1 \tag{6.75}$$

Solving for k, l, b gives

$$k = l = \frac{1}{\sqrt{1 - \left(\dfrac{v}{c}\right)^2}}, \quad b = \frac{v}{c^2} \qquad (6.76)$$

Let

$$\beta = \frac{v}{c}, \quad k = \frac{1}{\sqrt{1 - \beta^2}} \qquad (6.77)$$

Now, we have the transformation equations

$$t' = k\left(t - \frac{\beta}{c}x\right) \qquad\qquad t = k\left(t' + \frac{\beta}{c}x'\right)$$

$$x' = k(x - vt) \qquad\qquad x = k(x' + vt')$$

$$y' = y \qquad\qquad y = y' \qquad (6.78)$$

$$z' = z \qquad\qquad z = z'$$

These equations result in length contraction and time dilation. Consider a rod at rest in S' with its axis parallel to the X'-axis. Designating the two ends as a and b

$$x'_b = k(x_b - vt) \qquad (6.79)$$

and

$$x'_a = k(x_a - vt) \qquad (6.80)$$

If x_a and x_b of its two ends in S are determined at the same time then subtracting

$$x_b - x_a = \sqrt{1 - \beta}(x'_b - x'_a) \qquad (6.81)$$

Therefore, the measured length of the rod determined in moving system S' is less than that determined in rest system S' in the ratio

$$l_v = l_0 \sqrt{1 - \beta^2} \qquad (6.82)$$

Consider *a* clock at rest in *S'*. Let t'_a and t'_b be two times indicated by the clock. Then

$$
\begin{aligned}
t_b &= k\left(t'_b + \frac{\beta}{c} x' \right) \\
t_a &= k\left(t'_a + \frac{\beta}{c} x' \right)
\end{aligned}
\qquad (6.83)
$$

as the coordinates of the clock relative to the axes fixed in *S'* do not change with time. Subtracting

$$t_b - t_a = \frac{1}{\sqrt{1 - \beta^2}} \left(t'_b - t'_a \right) \qquad (6.84)$$

Th interval of time elapsed in S is greater than the interval in *S'*. Let t_a and t'_a be zero and let t_b be t_v and t'_b be t_0 then

$$t_v = \frac{t_0}{\sqrt{1 - \beta^2}} \qquad (6.85)$$

All this means is that if the speed of light can be measured accurately by instruments on moving frames then the length measured must shorten as given by (6.82) and the moving clock must run slower as given by (6.85). The Newtonian analyses of the recently discovered structure of matter show that matter behaves in this manner.

Einstein's analysis, based on the assumption that the speed of light relative to any reference frame was constant, gave length and time equations which were correct. Thus, his *principle of relativity* was apparently justified. In addition, and this is significant, no experiment, even now, has ever been inconsistent with his theory. Before the structure of matter was discovered, however, experiments on unstructured matter were remarkably inconsistent with Newtonian theory.

Summarizing now, it is safe to assume that an observer can use what he/she sees as though they were at rest with respect to the universal fixed reference frame. Scientists can use velocities they observe just as they were absolute velocities. At the present state of experimental accuracies, either the Newtonian system or the Einsteinian system can be used. The results will be indistinguishable, experimentally.

VII. CONCLUSIONS

We have shown that the Newtonian and Einsteinian theories of mechanics are experimentally indistinguishable. The primary reason for this is that the structure of matter is such that it shrinks as its velocity increases and orbit periods (times) increase with the velocity of matter. Matter growth, matter shortening, and time dilation are experimentally the same for both systems.

The Newtonian system uses a universal fixed reference system. The Einstein system uses an anthropic system—the reference system is fixed to the observer. The two systems produce the same results using current experimental accuracy. However, let us consider an imaginary experiment. We have a spaceship at rest. Observer A, a Newtonian weighing 200 kg, determines the mass of the spaceship as 10,000 kg. Observer B, an Einsteinian weighing 200 kg, also determines the spaceship mass as 10,000 kg. Observers A and B board the ship and accelerate to 0.866c. Observer A now weighs himself and the spaceship. He weighs 400 kg and the ship weighs 20,000 kg. Observer B, the Einsteinian, weighs himself and the ship. He still weighs 200 kg, and the ship only weighs 10,000 kg. The results are clearly distinguishable. It has been, and will be, difficult to construct an experiment which will distinguish between the two theories.

As the velocity of matter increases it contracts and orbital periods increase in such a way that in measuring the absolute speed of light the measurements give the absolute velocity accurately, no matter what speed the reference system has. Einstein mistakenly interpreted the result of the measurements meant that the speed of light was the same relative to any reference frame. Thus, with the speed of light being the same relative to any frame, he then assumed that the resulting matter contraction and time dilation was due to the way space and time were interconnected.

Einstein did not have a mechanism for mass growth. He simply postulated the mass growth with velocity based upon physical tests. The Compton scattering experiments clearly showed that mass was added to matter to accelerate it. It does not seem possible that an observer moving with an accelerated mass could remove the mass captured as a result of the acceleration.

The Einstein transformation equations deal with the dimensions of matter and matter clocks. The equations relate clock rates and matter lengths in one reference system to another reference system. He, and the world-wide physics community, believes the transformation equations relate space itself shrinking and a mysterious space clock running at a different rate from one reference system to another. Instead, the transformation equations account for the shortening of matter with velocity and the lengthening of time for an orbit in motion—as the matter moves.

VII. THE PARTICLE PROPERTY OF RADIATION

A. PHOTON STRUCTURE

The simplest hydrogen atom consists of an atom at rest where the atom consists of a proton and an electron orbiting the proton. A hydrogen atom can exist in many states. Another hydrogen atom can impact a resting hydrogen atom whose electron is in the closest orbit and transfer mass to the atom producing a translation of the atom. The atom is then in a higher energy state than before the impact. Before the transfer the mass to be transferred existed in the electrostatic *field* of the impactor. During impact a *string* of brutinos is formed having the shape of a harmonic wave with a length which would stretch around the impactor atom and then be transferred back into a near circular ring to the receiving atom. The result would be that the electron would take a larger orbit. The atom would be in a higher energy state. The transferred mass is a photon. The gross properties of the photon are derived in [7.1].

A photon is a cloud of brutinos stretched out along its complete wave length where the cross section is relatively constant along its length. Its amplitude is its wavelength divided by 2π. It translates at the speed of light, of course, and it does not undulate, but it retains its harmonic shape. Short wavelength photons have many brutinos for each brutino of length, and long wavelength photons have one brutino in many brutinos of length. The cloud is hardly perceptible from the ether waves which guide it, especially for long wavelength photons.

Photons make up light. They are a part of most of our energy sources; are involved in most, if not all, of our communication mechanisms; and make our bodies function. Photons are emitted by the Sun as well as by many stars. Photons are everywhere in the known universe. But, just what is a photon? As has been shown, the photon structure is much more precise than would be indicated by the gross characteristics derived in [7.3].

The short photons are $10^{-16}m$ long consisting of 10^{40} fundamental spherical ether particles and the long ones are 10^8m long consisting of 10^{16} ether particles. The short ones have 2×10^{21} continuous strings of particles stretching across their $10^{-16}m$ length. The long ones have 2×10^{16} particles spread out over the length of 10^8 meters so that only one part in 10^{27} of the string has particles.

The photon is produced and continually maintained by the electrostatic fields of the atom which emitted the photon, until it is transferred to other atoms. However, during transmission the photon loses one basic particle for each wave length of travel. Figure 7.1 illustrates the concept of photons. The ether particles are neatly organized as in the illustrations.

Ancient people knew light transmitted energy, could spread out spherically from a source, could be directed in a straight line ray, and that it appeared homogenous like a continuum of liquid.

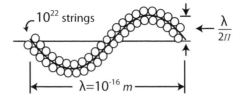

a. Short Fat Photon

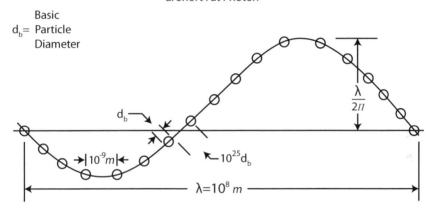

b. Long Skinny Photon

Figure 7.1. Long and Short Photons

Because of the fact that light could travel as a rectilinear ray without spreading appreciably, Sir Isaac Newton thought light was made up of tiny particles (which he called corpuscles).

In the years 1801-1804 Thomas Young produced an interference pattern between two light beams and explained the phenomenon using the wave theory of light. In fact, everything that was known about light up to the end of the 19th century could be explained by the wave theory — except for its rectilinear translation with very limited spreading.

Actually, Newton and Young were both correct. Light consists of corpuscles —the corpuscles are photons. A photon has a harmonic shape and many photons make up a continuum which acts as a wave.

Around 1860 James Clerk-Maxwell developed equations which unified electricity and magnetism. One consequence of his equations is for two electrons moving each with parallel and equal velocities v, where a line connecting the particles is perpendicular to the velocities, that the force between the electrons is $(e/r)^2-(v/c)^2(e/r)^2$ which tends to separate the electron. In this equation, e is the charge of the electron $(1.52\times10^{-14}kg^{1/2}m^{3/2}/s)$, r is the separation distance in meters, v is the velocity of the electrons in meters per second, and c is the speed of light $(3.00\times10^8\ m/s)$. The first term in the above expression is the electrostatic force of repulsion and the second term is the magnetic force of attraction. The force is measured in Newtons.

Another result of Maxwell's work is that the equations predicted the existence of electromagnetic waves which traveled at the speed of light. These predictions were verified experimentally by Hertz in 1887.

Newton's *mechanical* or *physical* corpuscles were abandoned in favor of some *etherial* compressible fluid for light which was the foundation of the wave theory. The postulated fluid had to be compressible in order to show the interference patterns.

Experiments on black body radiation, the photoelectric effect, Compton scattering, X-ray spectra, and optical spectra during the first two decades of the 20th century showed that light as well as all electromagnetic waves were made up of discrete particles which were named photons. Practically all photons are produced by atoms and many are absorbed by atoms.

The simplest case of a photon being emitted or absorbed by an atom involves a hydrogen atom. The simplest hydrogen atom consists of a proton and an electron orbiting as close as possible to the proton. A photon with a wavelength of $1.21\times10^{-7}m$ can change the electron orbit to the next higher energy state. Under certain circumstances, that hydrogen atom can emit a photon and the electron will return to its closest orbit.

As a result of experiments such as listed above, as well as other experiments, several characteristics of photons were determined. Every photon has an angular momentum of $h=6.6261\times10^{-34}$ kg-m^2/s, or an angular momentum per radian of $\hbar= h/(2\pi)=1.0546\times10^{-34}$ and a velocity in free space of $2.99792458\times10^8 m/s$. Photons occur in a wide spectrum of energies which are given in terms of their frequency as $E=h\nu$, where ν is the number of cycles per second which the photon produces as it translates. Photons also have a linear momentum, p, which is their energy divided by c. The wavelengths vary from 10^{-16} to 10^8 meters and beyond both limits.

Real insight into the structure of the photon came from our study of heat radiation including superconductivity (see Brown [7.1]). In order to be consistent with the experimental results of how heat escapes from matter, the photon had to consist of mass which was strung out uniformly along its complete wavelength.

An analysis of the angular momentum of an atom emitting a photon, based on the kinetic particle theory of the universe, shows that the photon must lose one ether particle for each wavelength of travel. This analysis means that the average photon can travel only to the edge of the observable universe, i.e., 10^{10} light years. This result is consistent with the *tired light* theory to explain the red shift of light from distant stars which was discovered by Hubble in 1929. We show this analysis in Section B of this chapter.

The first inkling that radiation consisted of particles came from a study of black body radiation. From these studies it became apparent that for a given temperature the radiation frequencies had a Maxwell-Boltzmann distribution, just as ideal gas particle velocities are distributed. Also the maximum probability frequency is directly proportional to energy of a *particle of radiation*. The maximum probability of a plot of radiation for a given temperature (i.e., particle energy) is a constant times the frequency. Further, the value of the constant which relates the frequency of a *radiation particle* to its energy is 6.63×10^{-34} *kg-m²/s*. This constant is the Planck constant, named after its discoverer Max Planck. It is signified by *h* and the *radiation particle* energy is *hv*, where *v* is the particle frequency in cycles per second.

To our knowledge, the structure described here is the only description of the photon structure anywhere in the literature of physics. It is difficult to conjure up the structure of something which added mass to matter and when it is emitted it becomes only energy (but with momentum) — as envisioned in the special theory of relativity.

B. THE NON-EXPANDING UNIVERSE

During the late 1960's our group at the McDonnell Douglas Corporation at the Santa Monica, Calif. Division was developing a theory of physics based upon a postulated kinetic particle universe. It was postulated that one type of particle, the brutino, made up everything in the universe. The background speed distribution of the brutinos is Maxwell-Boltzmann for which the ratio of the *RMS* speed v_r to the mean speed v_m is $v_r / v_m = \sqrt{3\pi / 8}$. Further, the numerical value of $[(v_r - v_m)/v_m]^2 = (\sqrt{3\pi / 8} - 1)^2 = 0.0729348$ was noted to be close to the fine structure constant; see the paper by Brown, Harmon, and Wood [5.2] and Chapter 2 of Brown [7.1]. Through the years members of the group discovered that $(v_r - v_m)$ is very close to the electromagnetic force velocity c, and that v_m is the strong nuclear force velocity; again see Chapter 2 of [7.1]. Thus, the square of the ratio $(v_r - v_m)/v_m$ was shown to be the ratio of the electromagnetic force to the strong nuclear force. Also, this ratio gave us the velocity of the brutinos since $v_r - v_m \approx c = (\sqrt{3\pi / 8})v_m - v_m)$. Thus $v_m \approx c/(\sqrt{3\pi / 8} - 1) \approx 2.998 \times 10^8 / 0.08540 = 3.510 \times 10^9$ m/s.

At the beginning of our research we did not know the brutino speed, mass, diameter, or mean free path. We did know that the (conventional) expanding universe theory was untenable with our postulates. Moreover, we knew that a photon must consist only of brutinos (since everything in the universe is postulated to be made up only of the kinetic particles called brutinos). Thus a photon energy must be the number of brutinos making up the photon times the mass of each times the square of their velocity, which is c^2. We assumed that the photon lost one brutino for each wave length of travel — the minimum it could lose. Now using an observable universe radius of 10^{25} meters, we determined that the brutino mass was in the order of 10^{-66} *kg* (see Chapter 6 of [7.1]), which is close to the magnitude we discovered by other methods (see Appendix A).

We propose the following mechanism. In the atom the photon predecessor has an angular momentum of h about the atom center (where h is Planck's constant). As the electron drops from one orbit to a lower orbit, the atom ejects a *string* of brutinos which makes a wave with half amplitude equal to the orbital radius where the stored photon mass resides and a length equal the orbital circumference. The atom proper loses an angular momentum of h but this angular momentum is maintained by the photon with the two electrical fields (i.e., the proton and electron) holding it together as it translates from the atom center. The angular momentum of the translating photon about the atom center is produced by the photon mass, i.e., brutinos in their harmonic shape as their distance from the atom increases. As the distance increases and since the velocity is constant (i.e., c), the photon mass must decrease in order to maintain the angular momentum at a constant value, i.e., h.

The following computations show the mass decrease. Let E_i be the photon energy after i wave lengths of travel from the atom. Now $E_i = M_i c^2 = hc/\lambda_i$ where M_i is the photon mass at the ith wave length from the atom, h is Planck's constant, and λ_i is the ith wave length. Also $E_{i+1} = M_{i+1} c^2 = hc/\lambda_{i+1}$. We have

$$E_i - E_{i+1} = M_i c^2 - M_{i+1} c^2 = (M_i - M_{i+1})\, c^2 = mc^2 \qquad (7.1)$$

where m is the mass of one brutino and it is assumed that the smallest amount of mass that can be lost will be lost. We also note

$$E_i - E_{i+1} = \frac{hc}{\lambda_i} - \frac{hc}{\lambda_{i+1}} = hc\left(\frac{cM_i}{h} - \frac{cM_{i+1}}{h} \right) = mc^2 \qquad (7.2)$$

The foregoing analysis shows a plausible mechanism for the *tired light theory* cause of the Hubble red shift. The plausibility of this theory should be compared with the plausibility of the expanding universe theory. Lastly, it is noted that the disturbance manifested as the photon is a result of the electron and proton fields being connected all the way to the edge of the observable universe.

Let us now compare this brutino theory with measurements. According to the expanding universe theory the velocity v of each distant star measured relative to the earth is directed from the earth to the star and its magnitude is

$$v=Hr \qquad (7.3)$$

where H is the Hubble Constant ($=1.80\times10^{-18}/s$), and r is the distance from Earth to the star. The velocity is related to the (observed) wavelength λ_{obs} by

$$v = Hr = c\frac{\lambda_{obs} - \lambda_o}{\lambda_o} \qquad (7.4)$$

where λ_o is the emitted wave length. The analysis given by this equation is limited to λ_{obs} being less than $2\lambda_o$, since a value greater than that would correspond to a star receding at a velocity greater than the speed of light. In that case no waves would reach the earth, according to the Hubble theory.

Let us now examine the consequences resulting from the indicated loss of one ether particle of mass m_b from a photon of mass M for each wave length of its travel. The amount of mass loss per meter of travel is

$$\frac{\Delta M}{\Delta r} = -\frac{m_b}{\lambda} \qquad (7.5)$$

where the negative sign indicates the reduction in mass and λ is the photon wavelength at the time the particle is removed. Initially m_b is relatively very small (30 orders of magnitude less than M) so that we can set $\Delta M / \Delta r$ equal to dM/dr. Now since $E=hc/\lambda=Mc^2$ we have

$$\frac{dM}{dr} = -\frac{m_b}{\lambda} = -\frac{m_b Mc}{h} \tag{7.6}$$

Integrating gives

$$\int_{m_o}^{m} \frac{dM}{M} = \frac{m_b c}{h} \int_0^r dr = \ln \frac{M}{M_o} = -\frac{m_b c}{h} r \tag{7.7}$$

Solving for r gives

$$r = -\frac{h}{m_b c} \ln \frac{M}{M_o} \tag{7.8}$$

Letting M_o be the mass of the highest energy photon emitted from a hydrogen atom we have

$$M_o c^2 = E_0 = E_2 - E_1 = 1/2 \alpha c^2 m(1/4 - 1) = -2.243 \times 10^{-16} J \tag{7.9}$$

Thus

$$M_o = 2.243 \times 10^{-16}/(3 \times 10^8)^2 = 2.492 \times 10^{-33} kg \tag{7.10}$$

Using $m = 2.89 \times 10^{-66}$ kg gives

$$r = \frac{1.054 \times 10^{-34}}{2.89 \times 10^{-66} \times 3 \times 10^8} \ln \frac{M}{2.492 \times 10^{-33}}$$

$$= -1.22 \times 10^{-23} \ln \frac{M}{2.492 \times 10^{-33}} \tag{7.11}$$

The following table gives values of r and wave length for several values of M/M_o.

$$(\lambda = (h/M_o)(M_o/M) = (1.054 \times 10^{-34})/(2.492 \times 10^{-33})(M_o/M)$$
$$= 0.0422 \ (M_o/M)) \tag{7.12}$$

M/M_o	0.9	0.5	0.1	0.0001	$m_b/M_o = 6.79 \times 10^{-34}$
$r - m$	1.28×10^{22}	8.45×10^{22}	2.81×10^{24}	1.13×10^{24}	8.93×10^{24}
$\lambda - m$	0.047	0.0844	4.22	422	6.22×10^{31}

The value of $r = 8.93 \times 10^{24}$ occurs when the photon is at its minimum mass and thus is the maximum range for the hydrogen emitted photon. However, near the end of the range, the analysis is not accurate due to the large percentage change in photon mass – as exemplified by the wave length of 10^{32} meters for the last photon wave. The value of the range computed (i.e., 8.93×10^{24}) compares favorably with the radius of the observable universe of 10^{10} *ly* since the radius in meters is

$$r_u = 10^{10} ly \times 365.25 d/y \ 24h/d \times 3600s/h \times 3 \times 10^8 m/s$$
$$= 9.47 \times 10^{25} m \tag{7.13}$$

Let us determine the values of range and velocity for small values of range, and then determine the predicted value of the Hubble constant, based on this theory. Let us use $M/M_o = 0.9$. For this value the range is 1.28×10^{22} m (see forgoing table) and the velocity based on assuming the wave shift is due to a receding source (i.e., a Doppler shift) rather than the mass loss is

$$v = c \frac{\lambda_{obs} - \lambda_o}{\lambda_o} = c \frac{\dfrac{h}{M_{obs}c} - \dfrac{h}{M_o c}}{\dfrac{h}{M_o c}} = c\left(\frac{1}{0.9} - 1\right) = 0.111c \tag{7.14}$$

Thus

$$H = \frac{v}{r} = \frac{0.111 \times 3 \times 10^8}{1.28 \times 10^{22}} = 2.60 \times 10^{-15} s^{-1} \tag{7.15}$$

This is over a thousand times larger than the measured value based on the Doppler shift theory, ($H_{Doppler} = 1.80 \times 10^{-18} s^{-1}$).

In this kinetic particle theory of the photon, the wave length shift can be tens of orders of magnitude larger than the emitted wave length (i.e., for a star at a distance having a velocity near the speed of light). Nonetheless, the predictions of the radius of the observable universe for both theories are somewhat consistent.

It is generally assumed that photons do not dissipate with travel. If so and if the universe is steady state, then the ambient temperature in the universe would be several million degrees. This high temperature would result from the radiation emitted by the stars throughout the universe. The expanding universe concept solved the temperature problem since stars running away from the earth, for instance, gave the same effect on the earth as dissipating photons. Our analysis of the problem results from our assumption that the emitted photon/emitting atom system angular momentum remains constant during emission and transmission.

C. THE HUMAN COSMOS

The human cosmos is the part of the universe which can be perceived by humans. It is the portion contained in a sphere with a radius of 10^{10} light years with its center at the earth.

Where did the sun come from? Where did the earth come from? Why was the earth initially hot and then became cool? What in the world are black holes and where do they come from? We present answers to these questions.

We normally think of the cosmos as the comets, planets, stars, galaxies — the whole unlimited universe. Here we consider that portion of the universe of which humans are aware, which we call the human cosmos. We consider humans as having existed in the universe for a very short time. The actual beginning may be some six thousand years ago according to the Judeo-Christian belief of human creation, or some hundreds of thousands of years ago according to the evolutionary belief of human creation – a short time in either case when compared with cosmological times. The extent in time and space of the cosmos of which humans are aware extends in time from the present back some ten billion years and extends outward in space some ten billion light years (or 10^{26} *meters*).

The theory of physics proposed in the book *The Grand Unified Theory of Physics* provides an approach to predicting the origin of the matter in the cosmos including the planets, stars, and even the *black holes*. We approach this with a theory which can produce all of the objects in the universe observed by humans. However there may be some material (such as stars and black holes) which was produced at some time long before the 10^{10} years and beyond distances much greater than 10^{26} meters from us and which migrated into the human cosmos space. In the following pages we describe the process for developing black holes, stars, and planets which appear in the human cosmos and which may, or may not, have been formed in the human cosmos.

In the grand unified theory of physics there is an ether gas which pervades the entire (assumed unlimited) universe. The gas is made up of elastic particles which have a diameter of $10^{-34}\,m$, a mass of $10^{-66}\,kg$, and an average speed of $3.5\times10^{9}m/s$ (approximately 10 times the speed of light). There are 10^{83} gas particles per cubic meter which gives a gas mass density of $10^{18}kg/m^{3}$ and a mean free path of $10^{-16}m$.

The most fundamental organization of these gaseous particles is a neutrino which translates at a speed equal to the background root mean square speed v_r less the mean speed v_m. This speed is very slightly greater than the speed of light. Thus the speed of light c is slightly less than $v_r - v_m$. Neutrinos are formed from the background with right-hand and left-hand angular momentum (about the translational axis) and in a spectrum of masses. The three different types of neutrinos (electron, muon, and tauon) are assumed to be identical except for their large variations of mass. Neutrinos are formed by the local complete condensation of the background gas and they occur in a large spectrum of mass due to the variation of mass of the condensed core of the neutrino. This condensation occurs inside a sphere whose radius is close to the background mean free path. The condensation produces a large thrust force which propels the neutrino through the background gas. The thrust force depends only upon the background properties and is independent of the neutrino mass. A proton is formed when a neutrino having a mass (i.e., energy divided by c^2) equal to the proton mass is knocked into a circular orbit by other massive neutrinos. Only one neutrino mass will balance the centrifugal force with the thrust force and produce the angular momentum which the neutrino had in its translational path, i.e. $\hbar/2$. That mass is the mass of the proton. The electron is formed concomitantly with the proton and, like the proton, it consists of one neutrino. The neutrino making up the electron presumably is formed by the flows of the proton. Thus, a hydrogen atom is formed by knocking one neutrino (having the mass of a proton) into a circular orbit. Antihydrogen is formed by knocking a left hand neutrino with the mass of a proton into a circular orbit.

Hydrogen atoms are continually formed by the process outlined in the above paragraph. Some, of course, are destroyed by the same process. In any case, there is a large number of hydrogen and antihydrogen atoms populating the entire universe and new ones are continually made. Large, long lived antihydrogen regions of the universe would have to be beyond our visible portion of the universe, i.e., beyond the range of our gravitational field.

A hydrogen atom has a gravitational field which extends 10^{26} meters in all directions. Due to this (gravitational) attractive force, and due to the random distribution of hydrogen atoms, the atoms begin congregating. This accumulation of hydrogen atoms continues until a hydrogen gas star is formed. As a star continues to grow it eventually reaches a mass large enough to produce atoms more massive than hydrogen. A neutron can be produced when the star's gravitational field is strong enough to accelerate a particle from space fast enough so that its energy can collapse the electron orbit in hydrogen.

Let us now determine the star mass required to accelerate a hydrogen atom located at a large distance to an impact energy equal to the orbital energy of an electron. We then will assume that energy approximates the energy required to produce a neutron or, more generally, fusion. The gravitational work to take a hydrogen atom from *infinity* to the star is

$$W=-\int_{\infty}^{R} \frac{GMm_h}{R^2} \, dR = \frac{GMm_h}{R} \qquad (7.16)$$

where m_h is the hydrogen mass and M is the star mass. The orbital kinetic energy of the electron, with mass m_e, in a hydrogen atom at rest is

$$E=\frac{1}{2}m_e(\alpha c)^2 \tag{7.17}$$

Using the star mass as

$$M=\frac{4\pi}{3}\rho R^3 \tag{7.18}$$

and equating the work to energy gives the star radius for fusion R_f as

$$R_f=\sqrt{\frac{3}{8\pi}\frac{m_e}{m_h}\frac{1}{\rho G}}\alpha c=\sqrt{\frac{3}{8\pi}\frac{1}{1836}\frac{1}{2960\times 6.67\times 10^{-11}}}$$
$$\times\frac{3\times 10^8}{137}=3.97\times 10^7\ m \tag{7.19}$$

The radius of the Sun is $1.6\times10^9 m$, forty times larger than this. Fusion, of course, continually occurs in the sun.

As a hydrogen star having a radius of $10^8 m$ continues to grow by collecting hydrogen atoms with its gravitational field it will eventually become a star of neutrons only, i.e. a neutron star. The minimum size of the beginning neutron star occurs when the gravitational force on a neutron is large enough to collapse the electron structure in neutrons. Equating the gravitational force to the centrifugal force of the electron we have

$$\frac{GM_n m_e}{R_n^2}=\frac{m_e(\alpha c)^2}{r_B} \tag{7.20}$$

where M_n is the star's mass, and its radius is R_n. From this

$$\frac{M_n}{R_n^2}=\frac{(\alpha c)^2}{Gr_B}=\frac{\left(3\times 10^8/137.1\right)^2}{6.7\times 10^{-11}\times 5\times 10^{-11}}=1.4\times 10^{33} \tag{7.21}$$

We use the mass of the neutron star just after it was formed. Thus, the density of the star is the density of the neutron. Now

$$\frac{M_n}{R_n^2} = \frac{\frac{4}{3}\pi R_n^3 \rho}{R_n^2} = \frac{4}{3}\pi R_n \frac{1.6\times10^{-27}}{\frac{4}{3}\pi\left(10^{-15}\right)^3} = 1.6\times10^{18}\,R_n \qquad (7.22)$$

From (7.21) and (7.22) we have

$$R_n = \frac{1.4\times10^{33}}{1.6\times10^{18}} \approx 10^{15}\,m \qquad (7.23)$$

Comparing this with the radius of the Sun, we have

$$\frac{R_n}{R_s} = \frac{10^{15}}{10^8} = 10^7 \qquad (7.24)$$

Thus R_n is 10 million times the Sun's radius. The radius $10^{15}m$ is the minimum radius for a neutron star which is formed starting with hydrogen atoms. (Other neutron stars are formed by the explosion of a much larger neutron star.)

As the neutron star continues to grow, it will eventually reach a size where the gravitational force will collapse the nuclear structure. For this to occur, the gravitational force on the neutron must be large enough to overcome the centrifugal force on the electron neutrino in the neutron. Equating the gravitational force to the centrifugal force gives

$$G\frac{M_N m_e}{R_N^2} = \frac{m_e c^2}{r_e'} \qquad (7.25)$$

where M_N is the neutron star having the size required to collapse the neutron, R_N is its radius, and r_e is the electron orbital radius in the neutron. The neutron mass is M_N and the proton orbital radius is r_e'. The star mass is its volume times the density of the neutron.

$$M_N = \frac{4}{3}\pi R_N^3 \left[1.6\times10^{-27} \Big/ \left(\frac{4}{3}\pi \times 10^{-15\times3} \right) \right] = 1.6\times10^{18}\,R_N^3 \qquad (7.26)$$

From (7.25) and (7.26)

$$G1.6\times10^{18}R_N = c^2 / r'_e \qquad (7.27)$$

or

$$R_N = \frac{\left(3\times10^8\right)^2}{6.67\times10^{-11}\times10^{-15}}\frac{1}{1.6\times10^{18}} = 8.43\times10^{23}\,m \qquad (7.28)$$

This is a large neuron star. Its radius is on the order of one millionth of the size of our visible cosmos.

When a neutron star reaches the size capable of collapsing the neutron structure, whose radius is $10^{15}m$, the neutrinos freed by the collapsing neutrons will cause the (giant) neutron star to explode. All kinds of matter will be thrown out.

A small chunk of debris could be a small-size mass of pure neutrons. Such a body as this would soon grow into a star as a result of the neutron disintegration and much of the mass would be retained in the form of the complete spectrum of atoms. The star would begin cooling and eventually become a *cool* planet, such as the earth. Larger chunks of the debris could form hydrogen stars with the characteristics of hydrogen stars formed by gravitationally collecting free hydrogen atoms. Very large chunks of the debris could be dark matter which is believed to populate our cosmos. It appears that such dark matter of *small* size could only come from the disintegration of an extremely large neutron star.

The explosion of such a large mass would populate the human cosmos and many other human cosmos-sized volumes throughout the universe adjacent to the human cosmos. Furthermore, such a massive gravitational mass as a $10^{23}m$ radius neutron star would collect all the matter within the reach of that star's gravitational field (i.e. a distance of some 10^{26} meters). It can thus be concluded that all the large hunks of mass that we observe probably came from this one explosion. This then gives us a method for estimating when the observable universe was formed.

First it must be recognized that in the massive ($10^{23}m$ radius) neutron star the highest energy isotopes of an atom would prevail. For example potassium 40 (^{40}K) would prevail over argon 40 (^{40}Ar). Next, the very short-lived fission transmutations would practically all be completed long before now, i.e. since the explosion 10^{10} years ago. Thus, we can use long-lived transmutation to estimate the time of the explosion.

We now consider a sample of matter and measure the amounts of ^{40}K and ^{40}Ar. We find the ratio of ^{40}Ar to ^{40}K is 9.3. The half-life for ^{40}K is $1.25\times10^9 y$. The number of ^{40}K atoms N_K at any point in time t is given by

$$N_K = N_{K_0}\, e^{-\lambda t} \tag{7.29}$$

where λ is the disintegration constant ($\lambda N = -dN/dt$). The number of argon atoms N_{Ar} is

$$N_{Ar} = N_{K_0} - N_K \tag{7.30}$$

Eliminating N_{K_0} by combining the two equations and solving for t gives

$$t = \frac{1}{\lambda}\ln\left(\frac{N_{Ar}}{N_K} + 1\right) \tag{7.31}$$

We need to determine λ in terms of the half-life. Let R be the decay rate

$$R = -\frac{dN}{dt} = \lambda N_{K_0}\, e^{-\lambda t} \tag{7.32}$$

or, the decay rate at any time is given in terms of the initial decay rate R_0 as

$$R = R_0 e^{-\lambda t} \tag{7.33}$$

Now if we substitute R from (7.34) into (7.33) and let t be zero we have

$$R_0 = \lambda N_{K_0} \tag{7.34}$$

Let τ be the half life, which is the time when both N and R are reduced to half their initial values, then from (7.33)

$$\frac{1}{2} R_0 = R_0 \, e^{-\lambda\tau} \tag{7.35}$$

or

$$\tau = \frac{\ln 2}{\lambda} \tag{7.36}$$

or

$$\lambda = \frac{\ln 2}{\lambda} \tag{7.37}$$

We can now solve (7.35) for the time at which the explosion occurred by using the measured value of the number of ^{40}Ar atoms per ^{40}K atoms as 9.3 and the half-life as $1.25 \times 10^9 y$.

$$t = \frac{1}{\lambda} \ln\left(\frac{N_{Ar}}{N_K} + 1\right) = \frac{\tau}{\ln 2} \ln(10.3+1) = \frac{1.25 \times 10^9}{\ln 2} \ln(11.3) \tag{7.38}$$

$$= 4.37 \times 10^9 \; years$$

Which agrees approximately with other computations of the origin of ordinary matter in the human cosmos.

The cosmological theory presented here seems to offer a reasonable explanation of how stars and planets are formed. The theory should be evaluated in comparison with other cosmological theories.

This type of cosmological theory provides a new theory of the demise of dinosaurs. Due to the continual collection of space mass, the earth is increasing its mass now and, presumably, its mass has increased from its beginning. The diameter of the Earth most likely did not increase uniformly with time. It could be that at the time of the dinosaurs' demise, a sudden increase of the Earth density occurred with a simultaneous increase of the gravitational force at the earth's surface. This massive compression would produce extensive radiation and debris around the earth, killing much of the living things on the earth. Many organisms would survive, but with the greatly increased gravitational force, all large dinosaurs would be destroyed and would not be regenerated. Also, the large flying dinosaurs could not fly with the greatly increased gravitational force. The currently *in vogue* explanation of dinosaur extinction being due to a massive meteorite impact does not explain why the dinosaurs did not return. Incidentally, the Earth's largest animal of all time, the giant blue whale, is alive and well today thanks to the small force pulling it toward the earth's center. The reduced force, of course, is due to the whale living in water.

The human cosmos has a volume of $(10^{10})^3 = 10^{30}$ m^3. Presumably, space along with its usual contents of stars, planets, etc; extends indefinitely with contents similar to the contents of our cosmos. Undoubtedly, there are star/planet systems very similar to the Sun/Earth systems. This would imply that there are other living organisms throughout the universe — unless life is controlled by some unknown spiritual constraints.

D. How We Get Our Energy

Every cubic meter of space has enough energy to power everything on the earth for several centuries. How do we get the meager supply we use? There are ten steps in the process.

1. Nature developed a micro-rocket pump which makes a vacuum, sucks in ether particles, and thus produces neutrinos.
2. One mass of neutrinos makes protons (and simultaneously electrons <u>and</u> simultaneously makes hydrogen atoms) continually.
3. Hydrogen atoms have gravitational fields. Thus hydrogen atoms congregate and make hydrogen stars.
4. Hydrogen stars grow and make larger atom stars.
5. Larger atom stars continue to grow and their gravitational fields get strong enough to break down the atomic structure and make neutron stars.
6. Neutron stars grow to many <u>lightyear</u> size and break down nuclear structure – then the big-big-big bang!
7. Big-big-big bang makes dark mass, smaller neutrino stars, hydrogen stars (suns), planets (earth), asteroids, comets, and space dust.
8. Earth evolves, makes anaerobic bacteria, makes atmosphere, makes aerobic bacteria, makes fungus, makes plants, and makes animals.
9. Organic material makes oil, gas, and coal. Also the sun continually dumps energy on the earth.
10. Humans mine coal, pump the oil and gas, and use solar cells, water power, and wind mills to meet our energy needs.

VIII. Quantum Electrodynamics

A. The Schrödinger Equation

When a photon impacts a free matter particle at rest, mass and momentum are imparted to the particle and it accelerates to velocity v. The center of mass of the imparted mass is captured at such a radius that the angular momentum imparted to the system is equal to Planck's constant h. The imparted mass and the impacted particle mass remain at a fixed distance from each other and they begin rotating about their common center of mass as the system center of mass translates in a straight line. This motion is manifested as a matter particle undulating as it translates. The wavelength of this undulation is $h/(mv)$, where m is the matter particle mass. The Schrödinger equation models the dynamics of the motion of the matter particle relative to a reference frame moving with the captured mass/matter particle system. Solution to the equation gives the velocity of the matter particle as a function of the location of the particle. We derive the Schrödinger equation by balancing the centrifugal forces against the centripetal forces. Thus, we show that the Schrödinger equation is a Newtonian equation.

The Schrödinger equation models the dynamics of matter particles when they translate. The equation was *discovered* in 1925 by the Austrian physicist, Dr. Erwin Schrödinger. *Discover* is used since neither he, nor any other scientist, has been able to derive the equation from basic principles during the 90 plus years since its discovery. The equation, along with Einstein's special theory of relativity, ushered in a whole new theory of physical science, namely *Modern Physics* consisting of *quantum* theory and *relativistic* theory. These theories both rejected classical, or Newtonian, theory. The equation has been used without fail by scientists throughout the world for almost a century. There can be no doubt about its capability for accurately predicting the behavior of nuclear particles.

The Schrödinger equation is a second order partial differential equation in two variables, space and time. The total energy of the translating particle is needed and physicists all over the world assume that the energy for a free translating particle is $mv^2/2$, where m is the mass of the particle and v is its velocity. They conclude that this is the energy since the work done on the particle by the accelerating medium is $mv^2/2$, i.e., the *kinetic* energy. The difficulty with this value of energy shows up in solving the Schrödinger equation. For the free translating particle the Schrödinger equation is solved by *separation of variables* giving two ordinary differential equations—a *time* equation and a *space* equation. The wavelength for the particle is the deBroglie wavelength. The wavelength from the time equation, using the total energy as $mv^2/2$, is twice the deBroglie length, which is impossible. From this we conclude that the total energy is mv^2.

Next, we examine the space equation. Using the energy as mv^2 gives the result that the wavelength from the space equation is half the deBroglie length. Replacing the factor 2 in the denominator of one term of the Schrödinger equation by unity results in a space equation with the same wavelength as the deBroglie wavelength.

From the above two paragraphs we have proven that the Schrödinger equation has two errors: the particle total energy is half the actual energy and the factor 2 should not be in the original equation. The two errors used in the space equation balance each other so that predictions of the equation are accurate. The equation is useful, without fail.

With our model of a translating matter particle we show

that the interpretation of the psi (ψ) function in the Schrödinger equation is not related to the probable location of the particle but is the velocity of the particle relative to a reference frame translating with the particle. We then show that the (corrected) Schrödinger equation results from summing forces, i.e., applying Newton's equations.

B. KINEMATICS OF THE MOTION OF A TRANSLATING PARTICLE

In the following discussion we consider the translation of an electron. Figure 8.1 shows a translating electron.

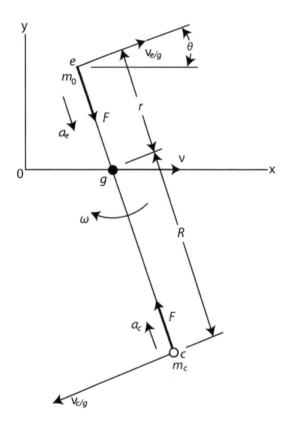

Figure 8.1. Geometry of the Translating Electron

The electron is labeled e in the figure. It rotates in a circle of radius r with velocity $v_{e/g}$ relative to the moving frame about the electron mass/captured mass center of gravity, g. The captured mass, m_c, rotates about g with a radius R and a velocity $v_{c/g}$ relative to the moving frame. The angular velocity is ω. The acceleration of the electron is a_e and the centripetal force is F. The acceleration of the captured mass, m_c, is a_c and its centripetal force also is F.

Figure 8.2 shows the absolute motion of the electron and captured mass as a function of distance traveled.

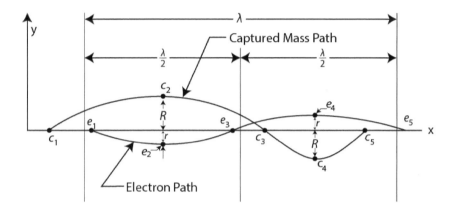

Figure 8.2. Paths of an Electron and Captured Mass

Figure 8.3 shows the path of the electron relative to the translating system. We note that the path is a circle.

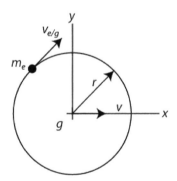

Figure 8.3. Electron Path Relative to the Electron/Captured Mass System

While the electron travels one (deBroglie) wavelength the electron rotates one cycle, or a distance of $2\pi r$. Thus, $2\pi r$ is its path length and it travels this distance while the electron travels λ, the deBroglie wavelength. Now

$$2\pi r = \lambda \tag{8.1}$$

Further, $v_{e/g}$ must be the same as v since the distances and elapsed times are the same.

The vertical and horizontal components of the velocity $v_{e/g}$ (= v) are harmonic functions, obviously. Figure 8.4 shows the horizontal component of the electron velocity relative to the translating reference system.

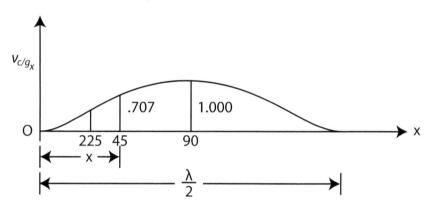

Figure 8.4. Horizontal Component of Electron Velocity vs. Location

This distribution is the same as obtained from the solution to the Schrödinger equation. However, quantum mechanicists incorrectly assume ψ is related to the probable dwell time.

$$p_{1-2} = A \int_{x_1}^{x_2} \psi\psi * dx \tag{8.2}$$

where p_{1-2} is the probability that the electron is located between distances x_1 and x_2. A is a *normalizing* constant which makes the total probability unity, and ψ^* is the complex conjugate of ψ.

The current quantum mechanics paradigm assumes that ψ is related to the *dwell time* of the electron along its path. We now determine the dwell time distribution based on our model of the translating electron.

From the equation for velocity

$$V_{e/g_x} = v_x = \frac{dx}{dt} \tag{8.3}$$

we have

$$dt = \frac{dx}{v_x} = \frac{dx}{v\sin\theta} \tag{8.4}$$

Then the time τ_{1-2} to move from x_1 and x_2 where the distances are measured relative to the moving frame is given by

$$\tau_{1-2} = \int_{x_1}^{x_2} \frac{dx}{v\sin\theta} = \int_{x_1}^{x_2} f_t\, dx \tag{8.5}$$

where f_t is defined as the dwell-time frequency and is given by

$$f_t = \frac{1}{v\sin(x/\lambda)} = \frac{1}{v\sin(2\pi x/\lambda)} \tag{8.6}$$

Now we can write

$$v\tau_{1-2} = \int_{x_1}^{x_2} \frac{dx}{\sin(x/\lambda)}, \quad vdt = \frac{dx}{\sin(x/\lambda)} \tag{8.7}$$

A plot of $f_t v$ is given in Figure 8.5.

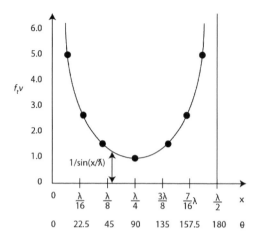

**Figure 8.5. Electron Dwell Time Frequency
Versus Distance Along Path**

Comparing Figure 8.5 with Figure 8.6, the Schrödinger solution, we conclude that the psi (ψ) function is actually the electron velocity. The real part and the complex part give the velocity components parallel to the x and y axes.

From the above we replace ψ by v in the Schrödinger equation. Now v is a complex number given by

$$v = v\sin\theta + iv\cos\theta \tag{8.8}$$

where $v\sin\theta$ is the horizontal component of the velocity and $iv\cos\theta$ is the vertical component.

C. CORRECTING THE SCHRÖDINGER EQUATION

The Schrödinger equation for the free translation of a particle of mass m is

$$\frac{\hbar^2}{2m}\frac{\partial^2\psi}{\partial x^2}+i\hbar\frac{\partial\psi}{\partial t} \tag{8.9}$$

where $\psi(x,t)$ is a function associated with the probable location of the particle. Assuming ψ is a function of x and t separately the equation is solved by using

$$\psi(x,t)= X(x)T(t) \tag{8.10}$$

where X is a function of x only and T is a function of t only. Substituting (8.11) into (8.10) gives the two ordinary differential equations

$$\frac{i\hbar}{T}\frac{dT}{dt}= E \tag{8.11}$$

and

$$\frac{\hbar^2}{2m}\frac{1}{X}\frac{d^2X}{dx^2}=-E \tag{8.12}$$

In these expressions E is the separation constant and is the total energy of the free particle.

Solving (8.12) gives

$$T(t)= \cos\frac{Et}{\hbar} -i\sin\frac{Et}{\hbar} \tag{8.13}$$

Quantum electrodynamacists use the total energy as $mv^2/2$. Substituting the energy into the arguments of cosine and sine in (8.13) and using $\hbar$ given by the deBroglie wavelength of the particle, i.e.,

$$\hbar = mvr \tag{8.14}$$

gives

$$\frac{Et}{\hbar} = \frac{\left(mv^2/2\right)t}{mvr} = \frac{v}{2r}t \tag{8.15}$$

The period of this wave, τ_t, is obtained by setting $vt/(2r)$ equal to 2π. Thus

$$\frac{v\tau_t}{2r} = 2\pi, \quad \tau_t = 2\frac{2\pi r}{v} \tag{8.16}$$

The period of the deBroglie wave, τ_d, is

$$\tau_d = \frac{2\pi r}{v} \tag{8.17}$$

From (8.16) and (8.17) the ratio of the period from the time equation to the deBroglie period is

$$\frac{\tau_t}{\tau_d} = \left(2\frac{2\pi r}{v}\right) \div \left(\frac{2\pi r}{v}\right) = 2 \tag{8.18}$$

which, of course, is impossible. The ratio of the wavelengths is

$$\lambda_t / \lambda_d = 2 \tag{8.19}$$

since the time is doubled and the velocity is the same. If in (8.15) we use the energy as mv^2 then (8.15) becomes

$$\frac{Et}{\hbar} = \frac{\left(mv^2\right)t}{mvr} = \frac{v}{r}t \tag{8.20}$$

The period τ_t now is given by

$$\frac{v\tau_t}{r} = 2\pi \, , \quad \tau_t = \frac{2\pi r}{v} \tag{8.21}$$

and

$$\frac{\tau_t}{\tau_d} = \left(\frac{2\pi r}{v}\right) \div \left(\frac{2\pi r}{v}\right) = 1 \tag{8.22}$$

From this we conclude that the particle total energy is mv^2.

Let us return to the space equation (8.12). Letting

$$e^2 = \frac{2mE}{\hbar^2} \tag{8.23}$$

We solve (8.12) by writing

$$\frac{d^2 X}{dx^2} + e^2 X = 0 \tag{8.24}$$

A solution is

$$X = A\sin(ex) + B\cos(ex) \tag{8.25}$$

which is easily seen to be a solution by using (8.23) and substituting (8.25) into (8.24).

$$-Ae^2 \sin(ex) - Be^2 \cos(ex) + e^2 A\sin(ex) + e^2 B\cos(ex) = 0 \quad (8.26)$$

The wavelength, λ_x, for the wave represented by this equation is obtained by setting the sine and cosine arguments equal to 2π. Thus

$$e\lambda_x = 2\pi = \sqrt{\frac{2mE}{\hbar^2}}\lambda_x \tag{8.27}$$

Using the incorrect energy $E=(mv^2)/2$ we have

$$\lambda_x = \frac{2\pi\hbar}{\sqrt{2mmv^2/2}} = \frac{mv\lambda_d}{mv} = \lambda_d \qquad (8.28)$$

Thus, the incorrect energy results in a space equation wavelength which agrees with the deBroglie equation. The reason for this agreement is that the Schrödinger equation is incorrect. Of course if in the term $\sqrt{2mmv^2/2}$ of (8.28) the first 2 were replaced by 1 and the second 2 also were replaced by 1 then the space equation would agree with the deBroglie wavelength. *Two* wrongs make a *right* in this case.

The Schrödinger equation (8.9) is corrected by removing the 2 in the denominator of the first term. In this case we let e have the value

$$e^2 = \frac{mE}{\hbar^2} \qquad (8.29)$$

and E has the value

$$E = mv^2 \qquad (8.30)$$

Now

$$e\lambda_x = 2\pi = \sqrt{\frac{mE}{\hbar}}\lambda_x = \frac{\sqrt{m^2v^2}}{\hbar}\lambda_x = \frac{mv}{mvr} = \frac{\lambda_x}{r} \qquad (8.31)$$

Further

$$2\pi r = \lambda_x = \lambda_d \qquad (8.32)$$

Using the corrected Schrödinger equation and the correct total energy results in a space equation wavelength the same as the deBroglie wavelength. The two must be the same.

D. Model of a Moving Matter Particle

To accelerate, a particle mass impacts the particle. Part of the mass scatters and part of the mass is captured by the *field* of the particle. The captured mass and the particle begin rotating about the combined mass center of the two masses as they translate. This motion is the undulation, or wave property, of translating particles. The center of mass of the captured mass is located at a position which results in the particle having an angular momentum of *h*. The resulting motion of the particle then has a *flow* component at velocity *v* (and an energy of $(1/2)mv^2$) plus a *thermal* component. The thermal component is the particle taking a circular path relative to the straight-line center of mass path. The circumference of this circular path is the same length as the wavelength so the velocity is *v*, same as the translational velocity. The thermal energy is, thus, also $1/2mv^2$. The total energy *E* then is

$$E = \frac{1}{2}mv^2 + \frac{1}{2}mv^2 = mv^2 \qquad (8.33)$$

We, of course, had discovered that the total energy had to be *mv*, from solving the Schrödinger equation.

Making the above correction we obtain the equation for a *free* particle as

$$\frac{\hbar^2}{m}\frac{\partial^2 \psi}{\partial x^2} + i\hbar\frac{\partial \psi}{\partial t} = 0 \qquad (8.34)$$

and the correct particle total energy is

$$E = mv^2 \qquad (8.35)$$

Now, recalling our discussion of the kinematics of the electron motion we replace ψ by the particle relative velocity, *v*, so that the corrected Schrödinger equation is

$$\frac{\hbar^2}{m}\frac{\partial^2 v}{\partial x^2} + i\hbar\frac{\partial v}{\partial t} = 0 \qquad (8.36)$$

An electron is accelerated by photons impacting it. The photons add momentum and mass to the particle. The mass added, m_c, is

$$m_c = \frac{m_o}{\sqrt{1-(v/c)^2}} - m_o \qquad (8.37)$$

where m_o is the matter particle mass when at rest, v is the particle velocity, and c is the speed of light.

The mass added, or *captured mass*, m_c, is assumed to continue moving at the velocity of the photon but in a closed loop around the electron. This is consistent with the energy of a moving matter particle having an energy of $m_o c^2/\sqrt{1-\beta^2}$. The electron of mass m_e and the captured mass m_c have a composite center of gravity g which translates in a straight path, of course.

The captured mass center c and the electron mass center e rotate around the center of gravity as the assembly translates. They are held together by electromagnetic forces. Figure 8.3 shows the two masses, m_o and m_c, their radii r and R, their velocities, $v_{e/g}$ and $v_{c/g}$, the force pair F and F holding them together, the accelerations a_c and a_e, the center of gravity g of the electron/captured mass, and the angular velocity ω. The electron (deBroglie) wavelength is $2\pi r$. While point g translates one wavelength, point e travels a circle relative to point g, which circle has a circumference of $2\pi r$. Thus, the velocity of e relative to g is v– same as the absolute velocity of g.

E. DERIVATION OF THE CORRECTED SCHRÖDINGER EQUATION

We can now show that the corrected Schrödinger equation is simply a balance of two Newtonian forces, i.e., the force pair holding the captured mass to the electron mass, as the masses translate and rotate about each other. Starting with equation (8.36) we divide by $\hbar$, multiply by m, let m_e be the electron mass, then the equation is

$$\hbar \frac{\partial^2 v}{\partial v^2} + im_e \frac{\partial v}{\partial t} = 0 \qquad (8.38)$$

where each term has the units of force. Next, we substitute for $\hbar$ from the deBroglie relation

$$\hbar = m_e vr \qquad (8.39)$$

Thus

$$m_e vr \frac{\partial^2 v}{\partial x^2} + im_e \frac{\partial v}{\partial t} = 0 \qquad (8.40)$$

The first term is the *space* term related to the electron mass and the second term is the *time* term related to the captured mass.

Using complex numbers, we can represent the velocity v as

$$v = v \cos \theta + iv \sin \theta \qquad (8.41)$$

This has the advantage of explicitly giving vertical and horizontal components for the parameters.

In the following analyses we need some partial differential relations. For the time equation,

$$\frac{\partial v}{\partial t} = \frac{\partial v}{\partial \theta} \frac{\partial \theta}{\partial t} = \frac{\partial v}{\partial \theta} \omega = \frac{\partial v}{\partial \theta} \frac{v}{r} \qquad (8.42)$$

For the space equation,

$$\frac{\partial^2 v}{\partial \theta^2} = \frac{\partial}{\partial \theta}\left(\frac{\partial v}{\partial \theta}\right) \tag{8.43}$$

and

$$\frac{\partial v}{\partial \theta} = \frac{\partial v}{\partial x}\frac{\partial x}{\partial \theta} \tag{8.44}$$

Further,

$$\frac{\partial x}{\partial \theta} = \frac{\partial x}{\partial t}\frac{\partial t}{\partial \theta} = \frac{v}{\omega} = \frac{vr}{v} = r \tag{8.45}$$

Now

$$\frac{\partial v}{\partial \theta} = \frac{\partial v}{\partial x}r = rf \tag{8.46}$$

where we have introduced

$$f = \frac{\partial v}{\partial x} \tag{8.47}$$

Returning to (8.43) and (8.46),

$$\frac{\partial^2 v}{\partial \theta^2} = \frac{\partial}{\partial \theta}\left(rf\right) = r\frac{\partial f}{\partial \theta} \tag{8.48}$$

where we have placed r to the left of the differentiation since r is constant. We can write the derivative of f as

$$\frac{\partial f}{\partial \theta} = \frac{\partial f}{\partial x}\frac{\partial x}{\partial \theta} = \frac{\partial f}{\partial x}r = \frac{\partial f}{\partial x}\frac{v}{\omega} = \frac{\partial f}{\partial x}\frac{v}{v}r = r\frac{\partial f}{\partial x} \tag{8.49}$$

Now, from (8.48), (8.49), and (8.47)

$$\frac{\partial^2 v}{\partial \theta^2} = r\left(r \frac{\partial f}{\partial x}\right) = r^2 \frac{\partial^2 v}{\partial x^2} \tag{8.50}$$

For the first term in (8.40) using (8.50) we have

$$m_e v r \frac{\partial^2 v}{\partial v^2} = m_e v r \frac{1}{r^2} \frac{\partial^2 v}{\partial \theta^2} = \frac{m_e v}{r} \frac{\partial^2}{\partial \theta^2} (v \cos\theta + iv \sin\theta)$$

$$= \frac{m_e v}{r} \frac{\partial}{\partial \theta}(-v\sin\theta - iv\cos\theta) \tag{8.51}$$

$$= \frac{m_e v}{r}(-v\sin\theta - iv\cos\theta) = -\frac{m_e v^2}{r}$$

For the second term in (10.40) we can replace m_c by $m_c R/r$ and v by $-v_{c/g} r/R$ to give the equation for the captured mass.

$$im_e \frac{\partial t}{\partial v} = -im_c \frac{R}{r} \frac{\partial v_{c/g}}{\partial t} \frac{r}{R} = -im_c \frac{\partial v_{c/g}}{\partial t} \tag{8.52}$$

The right-hand term in (8.52) is the captured mass times it's acceleration. Now using (8.41) the second term in (8.38) is

$$im_e \frac{\partial v_{c/g}}{\partial t} = -im_c \frac{v}{r} \frac{\partial v_{c/g}}{\partial \theta}$$

$$= -im_c \frac{v}{r} \frac{\partial \left(v_{c/g} \cos\theta + iv_{c/g} \sin\theta\right)}{\partial \theta}$$

$$-im_c \frac{v}{r}\left(-v_{c/g} \cos\theta + iv_{c/g} \sin\theta\right) \tag{8.53}$$

$$-im_c \frac{v}{r}\left(iv_{c/g} \cos\theta - iv_{c/g} \sin\theta\right)$$

$$= +im_e \frac{v}{r}\left(v_{c/g}\right) = -m_e \frac{r}{R} \frac{v}{r}\left(v\frac{R}{r}\right) = +\frac{m_e v^2}{r}$$

The corrected Schrödinger equation is

$$\frac{-m_e v^2}{r} + \frac{m_e v^2}{r} = 0 \tag{8.54}$$

Thus, we see that the Schrödinger equation, i.e., (8.38), consists of two terms: 1 the negative acceleration force acting on the electron and 2 the positive accelerating force on the captured mass. The Schrödinger equation is a Newtonian (classical) equation representing the dynamics of the motion of a translating particle.

Equation (8.36) is very similar to the Schrödinger equation for the free translation of a charged particle. The differences are that the Schrödinger equation has a factor 2 in the denominator of the first term and , presumably, is not a velocity. We have shown that the factor 2 is incorrect and, further, that ψ should be replaced by the particle (complex) velocity, v.

Einstein's theory of relativity and Schrödinger's equation of quantum mechanics gave birth to modern physics. The results of both have just been shown to be derived from Newtonian mechanics. Thus, *The Mechanical Theory of Everything* could signal the death of modern physics, and the revival of classical mechanics.

IX. Summary, Conclusions, and Recommendations.

We have investigated the principal problems in the physical sciences. We believe we have solved some of these problems and have given deep insight into many of the fundamental problems.

Our set of postulates upon which our theory rests is the simplest set imaginable. The postulates give a space for the *setting* of things in the universe, give a particle which introduces mass and three-dimensional geometry into our theory, give motion to introduce time, and give collisions to produce forces which cause change of motion.

The neutrino is the *organizer of the ether gas*. The neutrino takes particles from a Maxwell-Boltzmann homogeneous gas source, condenses them, and then aligns them so that a steady flow of the ether particles into the neutrino come out aligned in the same direction. In this process, less than half of the aligned particles exit from the front of the neutrino and the remainder exit at the rear at a lower speed, thus propelling the neutrino at the speed of the front exit velocity less the rear exit velocity — i.e., at a speed slightly above the speed of light.

Our research resulted in the discovery of the neutrino mechanism. The mechanism can occur inside a sphere having a radius approximately three times the mean free path. The mechanism *reduces the entropy in the universe* and is thus a counter-example to the second law of thermodynamics. If a *fluid sink* can be produced where ether particles coming into the sink reach sonic speed at a distance approximately three mean free paths from the center of the sink, then the ether gas can be condensed to the maximum density. In addition to our discovery of the significance of the phenomena occurring at the mean free path scale, we made another discovery. If particles are taken from a Maxwell-Boltzmann distributed gas, aligned so that their velocities are parallel and directed in the same sense without changing their speeds, and then squeezed together without changing their composite energy, then their transport velocity will increase by over eight percent. This mechanism propels the neutrino and provides the *vacuum pump* effect for bringing ether particles into the neutrino. The axial force produced by the pressure on the central part of the neutrino acts on a moving surface so that the force does work. The force is 1.43 meganewtons and is the fifth force of nature. It is the largest force, by far, of the five forces of nature. That work is the source of all usable energy in the universe. This phenomenon, to our knowledge, has not appeared in the scientific literature except for our writings. We call this pump the *nuclear pump*.

Now, by matching the *pump* size to the *sonic sphere* size, a stable inhomogeneous assembly of the (inert) ether particles can be constructed. Nature does such a construction and the result is the family of neutrinos.

Neutrinos are formed with a wide range of mass. The mass is mostly in the core which is very small. The angular momentum of a neutrino is mostly due to the changing of the flow direction while passing through the neutrino. The flow rate is the same for all neutrinos so that the measured angular momentum, $\hbar/2$, is the same for all neutrinos — independent of their mass. This constant flow rate also produces the same magnitude of thrust for all neutrinos.

We recommend that a *huge*, possibly a 10 meter diameter, spherical vacuum tank be constructed to model the neutrino using a gas (such as air) as the medium. Such an apparatus could produce a liquid or solid state of the gas at an extremely low rate. Most likely the rate would be too low for economic utility, but it could substantiate the neutrino theory. Further refinement of the experiment might yield a measurable force output.

A right-handed neutrino with a mass equal to the proton mass infrequently is knocked into a circular orbit with its thrust directed toward the center of the circle. When the resulting orbital radius has a value around $10^{-16}m$ the neutrino mass times this radius times the neutrino initial (and final) velocity (i.e., the speed of light) will be $\hbar/2$. The orbital path will be stable and the initial translating neutrino will become a proton. The orbiting neutrino then produces flow making the electron — which orbits the proton and, thus, produces the hydrogen atom.

The circular motion of the proton neutrino produces a magnetic moment and the electrostatic field. The distant motion of the ether is similar to that produced by a breathing sphere. However, the right-handedness of the neutrino produces polarity of the field. An anti-proton is produced by a left-handed neutrino. The circular motion of the proton also produces *close-in* wake flows which many physicists believe are part of the proton itself.

The output from the orbiting neutrino making the proton also produces *wavespaces* having their three dimensions commensurate with the proton orbital radius, i.e., the wavespaces are $10^{-16}m \times 10^{-16}m \times 10^{-16}m$. The wavespaces are formed at the periphery of the proton's orbit. They advance at the speed of light, $\approx v_r - v_m$, radially from the proton. These wavespaces are the *fine structure* of the proton's electrostatic field.

Simultaneously, with the formation of the proton, the electron is formed from the ether to counterbalance the positive electrostatic field of the proton. All matter at rest is made up of neutrinos taking a circular path. The electron neutrino has a mass $1/1836$th that of the proton or $1/(6\pi^5)$ times the proton mass. Thus, the electron neutrino orbit has an orbital radius $1/1836^{th}$ (or $\approx 10^{-19}m$). This path is called the *inertial* path. The electron neutrino makes a larger diameter $10^{-17}m$ loop to produce the electrostatic field and then a final circular loop (diameter $= 10^{-13}m$) to produce its angular momentum. The neutrino making the electron is left-handed which produces the negative charge. Our thinking is that the electron may be produced from the smallest possible neutrino.

The neutron is made from a hydrogen atom where the electron orbit center is forced down to the center of the proton.

The strong nuclear force between a proton and a neutron is the result of the two proton-sized neutrinos orbiting side-by-side which produces a flow velocity between them equal to the background mean speed. It is recommended that this model of the strong nuclear force be developed to predict, with precision, nuclear forces for all the atoms.

The electrostatic force between two charged particles is due to the *breathing sphere with twist* mechanisms of the two interacting particles. An electron orbiting a proton gives a higher order breathing sphere effect at a distance from the hydrogen atom which has an amplitude equal the brutino radius and this is the mechanism of gravitation. The analysis proving the mechanism of electrostatics and gravitation should be *revisited* especially in light of the recent determination of the constants of physics as well as the recent discovery of the fine structure.

The weak nuclear force, the fifth force of physics, is a decay force, i.e., it permits fundamental particles of physics to tear themselves apart. The basic mechanism of the weak nuclear force is due to two elementary matter particles rotating in various directions. The mechanism is illustrated by the neutron decay. The two orbiting neutrinos in the neutron put out four narrow *streams* of particles. These streams can be directed anywhere over 4π sterradians of direction. When the proton *output stream* happens to hit the orbiting electron squarely enough it can drive the electron to the Bohr radius and make a free electron and a free proton. The weak nuclear force is not strong enough to prevent decay.

Matter is accelerated by impacting it with photons. Mass of each photon is partially captured by the matter particle's electrostatic field and partially scattered. If the matter particle is initially at rest its neutrino path is circular. After impact, neither the neutrino mass[1] nor its speed is changed; it simply takes a plane spiral path. From a frame moving with the accelerated particles the path is an ellipse with its minor axis shorter than its radius when at rest. Thus, when accelerated from rest, matter's mass grows, its length shortens, and it takes a longer time to make an orbit.

The analysis here clearly shows that Newtonian mechanics is applicable at high speeds, which obviates the need of the Einstein theory of relativity. Further analyses of moving magnets and speed of light measurements show that the Einstein theory is unnecessary. Finally, it is expected that measurements of mass accelerated in different directions at the same place and time would clearly show the Einstein theory to be incorrect. Measurements of mass growth versus velocity on Earth (which may have an absolute velocity of several percent of the speed of light) should show different mass growths when accelerated in different directions on Earth. Just a few tests might establish the Earth's absolute velocity. It is recommended that these tests be performed <u>now</u>.

When matter is accelerated, the accelerating mass is captured off-center so that mass undulates as it translates. Their undulation makes the particle act as a *wave*. The undulating motion of a translating charged particle acts like a *breathing sphere* and produces the magnetic field.

A translating electron with a wavelength equal to the orbital circumference in a hydrogen atom produces a stable state. The

---

1 The *field* of the matter particle captures the impinging mass, not the neutrino.

electron's orbital velocity in speed of light units is the fine structure constant α ($= v/c = e^2/(\hbar c)$). The orbital angular momentum is $\hbar$ which results from balancing the electrostatic force against the centrifugal force and using the electron wavelength as the orbital circumference.

When a charged particle is accelerated its mass is increased. That mass is captured by the charged particle at a distance from the particle and it remains at that distance. The captured mass and the particles then begin rotating about their common mass center as they translate. The capture distance is such that the angular momentum of the system is increased by h, Planck's constant. As a result, the particle undulates as it travels. The resulting wave length then is $h/(mv)$, Planck's constant divided by the particle linear momentum. This, of course, is the wave length postulated by de Broglie. The particle energy as it translates then is that due to the linear motion $mv^2/2$ and the same amount due to the rotation. Thus, the total energy is mv^2. The dynamics of the motion is governed by Newton's equation of motion and produces an equation similar to the Schrödinger equation. We conclude that the corrected Schrödinger equation, which we derive, accurately models the motion of charged particles. We also make the more general conclusion that quantum electrodynamics is a classical phenomenon.

Cosmic rays are produced continually by protons and antiprotons which are made continually from the ether gas.

In conclusion we feel this book has presented significant inroads into understanding the physical world. Much work remains for a complete, rigorous understanding of the physical world.

As an afterthought, this theory really got started when it was realized that all nuclear matter consists of mass moving in a circle at the speed of light so that the energy of matter is given by $E = mc^2$, the famous formula discovered by Einstein.

And one more afterthought! Science only deals with changes.

Here we begin with a three-dimensional space, a brutino gas with the proper mean free path length to particle diameter ratio, and a certain average speed. We have no way of answering questions like:

> Why is space three-dimensional?
> Where did brutinos come from?
> Why is their average speed 3×10^9 *m/s*?

These may be questions for philosophers.

APPENDIX A.
THE BASIC CONSTANTS OF THE KINETIC PARTICLE THEORY OF PHYSICS

Four basic independent constants are required to define the kinetic particle theory of physics. The four we determine here are the mass of the basic particle (the brutino), m_b; the mean velocity of the particles, v_m; the radius of the basic particle, r_b; and the mean free path, ℓ.

The parameters m_b, r_b, and v_m give mass, length, and time to the theory. The mean free path gives a scaling to the theory. Two universes with the same values for m_b, r_b, and v_m could be quite different if they had different values for the mean free path.

We begin with the constants we know most accurately. From our analyses of the formation of photons and the mechanism of the propagation of neutrinos, we know that the speed of light, c, is slightly less than $v_r - v_m$. Further, from our analysis of the fine structure constant and its measured value, it appears that the speed of light is $0.999720879(v_r - v_m)$. From this and the fact that

$$v_r/v_m = \sqrt{3\pi/8} \qquad (A\text{-}1)$$

and, further, that the value of c is

$$c = 2.9979458 \times 10^8 m/s \qquad (A\text{-}2)$$

we have

$$c = 0.999720879(v_r - v_m) = 0.999720879[v_m \sqrt{3\pi/8} - v_m] \quad \text{(A-3)}$$

and

$$\begin{aligned}
v_m &= c/[0.999720879(\sqrt{3\pi/8} - 1)] \\
&= 2.9979458 \times 10^8/(0.999720879 \times 0.085401882) \quad \text{(A-4)} \\
&= 3.511353067 \times 10^8 m/s
\end{aligned}$$

From the kinetic particle theory of gravitation we obtained the expression that the brutino's radius is the Planck length, i.e.,

$$\begin{aligned}
r_b &= \sqrt{Gh/c^3} \\
&= \sqrt{\frac{6.673 \times 10^{-11} \times 2\pi \times 1.0546 \times 10^{-34}}{(2.9979 \times 10^8)^3}} \quad \text{(A-5)} \\
&= 4.052 \times 10^{-35} m.
\end{aligned}$$

This value of the particle radius is assumed to be as accurate as the square root of the gravitational constant is known. We assume this notwithstanding that the above equation for r_b obtained from our theory of gravitation might vary somewhat from the above equation.

The orbital radius of the proton is known as accurately as the lesser accuracy of the Planck constant or the proton mass. Now

$$\begin{aligned}
r_p &= \hbar/2m_p c \\
&= 1.0546 \times 10^{-34}/2(1.6726 \times 10^{-27} \times 2.9979 \times 10^8) \quad \text{(A-6)} \\
&= 1.0516 \times 10^{-16} m
\end{aligned}$$

The mean free path

$$\ell = 2.35 \times 10^{-16} m \qquad\qquad \text{(A-7)}$$

was determined by a long circuitous route. We used nucleon sizes and the strong nuclear force to determine the background mass density. Knowing the brutino mass, the particle number density, η_o, was determined. With η_o and the brutino radius known we were able to compute the mean free path.

Incidentally the thrust of the neutrino is known accurately – since it is assumed equal to the proton neutrino centrifugal force, thus the neutrino thrust T_v is

$$T_v = \frac{mv^2}{r} = \frac{m_p c^2}{r_p}$$

$$= \frac{(1.6726 \times 10^{-27})(2.9979 \times 10^8)^2}{1.0516 \times 10^{-16}} \qquad\qquad \text{(A-8)}$$

$$= 1.42947 \times 10^6 \ N$$

The mass of the brutino was estimated from the Hubble redshift analysis and is

$$m_b = 2.89 \times 10^{-66} kg \qquad\qquad \text{(A-9)}$$

Summarizing we have the four basic constants

$$v_m = 3.511353067 \times 10^9 \ m/s$$
$$r_b = 4.052 \times 10^{-35} \ m \qquad\qquad \text{(A-10)}$$
$$\ell = 2.35 \times 10^{-16} \ m$$
$$m_b = 2.89 \times 10^{-66} \ kg$$

Also

$$\eta_0 = \frac{1}{4\sqrt{2}\pi r_b^2 \ell} = 1.46\times10^{83}\,/m^3$$

$$\rho_o = 4.23\times10^{17}\,kg/m^3$$
$$v_r = 3.81124254\times10^9\,m/s \qquad\qquad (A\text{-}11)$$
$$r_c = 7.50\times10^{-16}\,m$$
$$s = 1/\eta_o^{1/3} = 1.899\times10^{-28}\,m$$

Appendix B.
Energy, Kinetic Energy, and Work

The energy of matter is mc^2 where m is the mass of the matter and c is the velocity of light. We consider an example of elastic balls accelerating a point mass m from zero to velocity v. The energy given up by the accelerating system is mv^2. We also show that the work done by the accelerating system is half the energy given up by the accelerating system. We then consider a matter particle being accelerated by photons. Here again, the energy required for accelerating a matter particle is greater than the work (and kinetic energy) applied to the matter particle.

We know that the energy of matter is mc^2, i.e., its mass times the square of velocity. We say that a ball of mass m translating at velocity v has a kinetic energy of $1/2\ mv^2$. But, isn't its energy mv^2? We think so. How do we reconcile the fact that the work, i.e. energy expended, which is $\int F\,dx$ required to bring the energy of a particle up to mv^2, when only half that amount of work is required?

Let us illustrate our dilemma with the following example. Let F be constant. Now we write Newton's equation, assuming non-relativistic velocities

$$Fx_f = \int_0^{x_f} ma\,dx = m\int \frac{dv}{dt}\,dx = m\int \frac{dv}{dx}\frac{dx}{dt}\,dx$$
$$= m\int_0^{v_f} v\,dv = \frac{mv_f^2}{2}$$

(B.1)

where x_f is the accelerating distance, m is the ball mass, a is the acceleration, v is the ball velocity, and v_f is the velocity at the end of the acceleration. The energy input is not equal to the ball energy change—it is only half the change of the ball's energy. How can that be? Is the conservation of energy law violated? Let us look at a more complete description of the acceleration process.

To examine the energy involved in the accelerating process, let us consider the acceleration of one elastic ball produced by the repeated impacts by other identical elastic balls. Let us number the impacts from 1 to N and let their velocities be $v_1, v_2, v_3, \ldots, v_n$. The individual masses of the impactors and impactees are all m. Let the impactee initially be at rest. Further, let the impact velocities all be central impacts and let all velocities be parallel to the impactee's velocity.[1] Consider the first impact by particle 1. The velocity v_1 is simply transferred to the impactee and the impactor's post-impact velocity is zero. The first impact removes energy mv_1^2 and transfers it to the impactee. The momentum added to the impactee is mv_1. Let the second impactor have velocity $2v_1$. Its post-impact velocity will be v_1 and the impactee's velocity will be $2v_1$. The momentum imparted will be mv_1, the pre-impact energy of the impactor will be $m(2v_1)^2$, and the post impact energy will be mv_1^2, or a loss of $4mv_1^2 - mv_1^2 = 3mv_1^2$. The pre-impact energy of the impactee was mv_1^2 and the post-impact energy of the impactee was $m(2v_1)^2 = 4mv_1^2$. The impactee received an energy of $3mv_1^2$ in this impact. Let us summarize the result of impacts.

---

[1] This can be possible only if the impactor is swept out of the approach direction of the impactors as soon as the impact is complete.

Impact Num.	Impactors		Impactee		Momen-tum Imparted
	Vel. Bef.	Vel. Aft.	Vel. Bef.	Vel. Aft	
1	v_1	0	0	v_1	mv_1
2	$2v_1$	v_1	v_1	$2v_1$	mv_1
3	$3v_1$	$2v_1$	$2v_1$	$3v_1$	mv_1
4	$4v_1$	$3v_1$	$3v_1$	$4v_1$	mv_1
...	...	...	...	...	...
N	Nv_1	$(N-1)v_1$	$(N-1)v_1$	Nv_1	mv_1

Impact Num.	Impactor Energy		Impactee Energy	
	Loss	Cumul.	Gain	Cumul.
1	mv_1^2	mv_1^2	mv_1^2	mv_1^2
2	$3mv_1^2$	$4mv_1^2$	$3mv_1^2$	$4mv_1^2$
3	$5mv_1^2$	$9mv_1^2$	$5mv_1^2$	$9mv_1^2$
4	$7mv_1^2$	$16mv_1^2$	$7mv_1^2$	$16mv_1^2$
...	...	...	...	...
N	$(2N-1)mv_1^2$	$N^2mv_1^2$	$(2N-1)mv_1^2$	$N^2mv_1^2$

We have the impact velocities increase for each subsequent impact by an integral amount of the first impact. Further, the spacing increases so that the impacts occur at equal intervals. Thus, the force applied to the impactees will be constant. Now

$$\tau = \frac{l_n}{v_n} \tag{B.2}$$

and

$$l_n = nl_1, \qquad v_n = nv_1 \tag{B.3}$$

$$\tau = \frac{l_n}{v_n} = \frac{nl_1}{nv_1} = \frac{l_1}{v_1} \qquad \text{(B.4)}$$

We see that this makes the impacts occur at equal time intervals. Force is the time rate of momentum imparted. Thus

$$F = \frac{mv_1}{\tau} = \frac{mv_1 v_1}{l_1} = \frac{mv_1^2}{l_1} \qquad \text{(B.5)}$$

We see the force is constant, mv_1^2/ℓ. The impactor system energy decreases by $N^2 mv_1^2$ and the impactee energy increases by the same amount. Thus, energy is conserved in this acceleration process. The work W done is the force times the distance, which is

$$
\begin{aligned}
W_N = F \times (dist) &= \frac{mv_1^2}{l_1}(l_1 + l_2 + l_3 + ... + l_N) \\
&= \frac{mv_1^2}{l_1}(l_1 + 2l_1 + 3l_1 + ... + Nl_1) \\
&= mv_1^2(1 + 2 + 3 + ... + N) \\
&= mv_1^2 \frac{N(N+1)}{2}
\end{aligned}
\qquad \text{(B.6)}
$$

Note that

$$E_N = mv_1^2 N^2 \qquad \text{(B.7)}$$

Thus

$$\frac{E_N}{W_N} = \frac{mv_1^2 N^2}{mv_1^2 N(N+1)} \approx 2 \qquad \text{(B.8)}$$

For large N the energy expended is twice the work produced. Similarly the energy received is the same as the energy expended, but twice the work (energy) done on the accelerated particle. From thus we have

$$Work\ Applied = \frac{Energy Added}{2} = Kinetic\ Energy \qquad (B.9)$$

Energy is mass times velocity squared, and energy is conserved in this example.

The foregoing analysis considers the acceleration of a point mass. Let us now consider the acceleration of a matter particle impacted by photons. In this case the photon mass is partially captured and partly captured. The momentum imparted, averaged for many impacts, is the total photon momentum. The captured mass still moves at the velocity of light but in a closed loop around the initial matter particle. The *loop* is seen as a plane spiral from the initial frame but from a frame moving with the matter particle it is seen as an ellipse. Similarly, the neutrino making the matter particle takes a plane spiral path which is an ellipse when seen from a frame moving with the particle.

For the analysis of the acceleration of a matter particle, we apply a force F while the particle translates a distance x_f. For simplicity we consider photons with momentum mc applied at equal time intervals at $\dot{n}$ photons per second. The work done then is

$$W = Fx_f = \int mc\dot{n}\,dx \qquad (B.10)$$

This force produces the acceleration

$$F = ma = \frac{m_o}{\sqrt{1-\beta^2}}\frac{d^2x}{dt^2} = \frac{m_o}{\sqrt{1-\beta^2}}\frac{dv}{dt} \qquad (B.11)$$

The work done is

$$W = \int F\,dx = \int \frac{m_o}{\sqrt{1-\beta^2}}\frac{dv}{dt}dx = \int \frac{m_o}{\sqrt{1-\beta^2}}v\,dv \qquad (B.12)$$

Integrating this gives

$$W = \int \frac{m_o}{\sqrt{1-\beta^2}} v\, dv = c^2 \int \frac{m_o}{\sqrt{1-\beta^2}} \beta\, d\beta$$

$$d\left(\sqrt{1-\beta^2}\right) = \frac{1}{2} \frac{-2\beta d\beta}{\sqrt{1-\beta^2}} = \frac{-\beta d\beta}{\sqrt{1-\beta^2}} \tag{B.13}$$

$$\beta d\beta = -\sqrt{1-\beta^2}\, d\left(\sqrt{1-\beta^2}\right)$$

Now

$$W = -m_o c^2 \int_0^{\beta_f} \frac{\sqrt{1-\beta^2}\, d\left(\sqrt{1-\beta^2}\right)}{\sqrt{1-\beta^2}}$$

$$= -m_o c^2 \left[\sqrt{1-\beta^2} \right]_0^{\beta_f} = -m_o c^2 \left[\sqrt{1-\beta_f^2} - 1 \right] \tag{B.14}$$

$$= m_o c^2 \left[1 - \sqrt{1-\beta_f^2} \right]$$

For small β_f

$$W = m_o c^2 \frac{1}{2} \beta_f^2 = \frac{1}{2} m_o v_f^2 \tag{B.15}$$

For large β_f

$$W = m_o c^2 \tag{B.16}$$

The energy E of the particle is

$$E = \frac{m_o c^2}{\sqrt{1-\beta_f^2}} \tag{B.17}$$

The ratio of the energy to the work is

$$\frac{E}{W} = \frac{m_o c^2}{\sqrt{1-\beta_f^2}\, m_o c^2 \left[1-\sqrt{1-\beta_f^2}\right]} = \frac{1}{\sqrt{1-\beta_f^2}-1+\beta_f^2} \quad (B.18)$$

For large β_f

$$\frac{E}{W} = \infty \quad (B.19)$$

For small β_f

$$\frac{E}{W} = \frac{1}{1-\frac{1}{2}\beta_f^2 - 1 + \beta_f^2} = 2\beta_f^2 \quad (B.20)$$

We now obtain the relation between energy and momentum. Start with (B.17). Square the expression and divide by $m_o^2 c^4$.

$$\frac{E^2}{m_o^2 c^4} = \frac{1}{1-\beta^2} = 1 + \frac{\beta^2}{1-\beta^2} \quad (B.21)$$

Let us square the momentum $p(= mc)$ and divide by $m_o^2 c^4$

$$\frac{p^2}{m_o^2 c^4} = \frac{v^2}{\left(1-\beta^2\right)c^4} = \frac{\beta^2}{\left(1-\beta^2\right)c^2} \quad (B.22)$$

Substituting (B.21) into (B.22) gives

$$\frac{E^2}{m_o^2 c^4} = 1 + c^2 \frac{p^2}{m_o^2 c^4} \quad (B.23)$$

Multiplying by $m_o^2 c^4$ gives

$$E^2 = m_o^2 c^4 + c^2 p^2 \quad (B.24)$$

REFERENCES

2.1 Kennard, Earle H., *Kinetic Theory of Gases*, Chapter II, McGraw-Hill Book Company, Inc. New York, 1938.

3.1 Binder, R.C., *Fluid Mechanics,* 3rd ed., Prentice-Hall, Inc., Englewood Cliffs, N.J., 1955.

4.1 Bassett, A.B., *A Treatise on Hydrodynamics,* Vol. 1, Dover Publications, New York, 1961.

4.2 Whittaker, E.T., *A History of the Theories of Aether and Electricity I*, Pages 284ff. Thomas Nelson and Sons Ltd., London, 1951.

4.3 Brown, J.M., "Force Production from Interacting Gas Flows for BMD Applications", Final Report on U. S. Army Contract DAS6-80-C-0034 Administered by U. S. Army Ballistic Missile Defense Agency, Box 1500, Huntsville, Al. 35807, October 1, 1981.

4.4 Brown, Joseph M., *Principles of Science*, ISBN 0-9626768-0-2, Basic Research Press, 120   East Main Street, Starkville, MS 39759, 1991.

5.1 Mohr, P.J.; Taylor, B.N., and Newell, D.B., (2015). *The 2014 CODATA Recommended Values of the Fundamental Physical Constants.* (Web Version 7.0). http://physics.nist.gov/constants

5.2 Brown, J.M., Harmon Jr., D.B., and Wood, R.M., "A Note on the Fine Structure Constant," McDonnell Douglas Astronautics Company Paper MDAC WD 1372 Huntington Beach, CA, June 1970.

5.3 Cham, Jorge and Whiteson, Daniel; *We Have No Idea* ISBN 978-0-7352115-1-3, Riverhead Books. New York, NY. 2017.

6.1 McClusky, S.W., *Introduction to Celestial Mechanics*, Addison-Wesley Publishing Company, Inc., Reading Massachusetts, 1963.

6.2 Page, Leigh; *Introduction to Theoretical Physics, 3rd ed.*; Van Nostrand Company, Inc.; Princeton, NJ, 1952.

7.1 Brown, Joseph M., *Photons and the Elementary Particles*, ISBN: 0-978-0-9712944-5-5, Basic Research Press, 120 East Main Street, Starkville, MS 39759, 2011.

INDEX

A

absolute velocity of the earth
Aerospace vii
Aerospace Corporation vii
Aguilar-Benite
angular momentum , 33, 43, 47,
    49, 50, 61, 79, 81, 85, 150,
    113, 116, 152, 154, 24
anti-hydrogen 67
anti-matter 67
antimatter 54
anti-proton 151
aware 121

B

Ballistic Missile Agency viii
basic constants vii, 34
Bassett 43
behavior
big-bang xix
Binder, R.C.
Bjerknes viii
Bjornlie vii
black holes 121
blue whale 129
Bohr 49, 153
Boltzmann viii, 7, 8, 113, 115,
    149, 150
breathing sphere ix, 43, 44, 45,
    64, 65, 151, 152, 152, 153
Brown

Brown, J.M. , 167, 168
brutino xx, 3, 4, 6, 8, 64, 64, ,
    109, 152, 115, 116, 23, 53,
    54, 53

C

Cancer xvi
cell xvi
center of charge 81
center of mass 78, 81, 83
center-of-mass 78
Christian 121
complex 20
compression chamber 20, 21
Compton
condensation 122
control vii, 38
converging nozzle 25
corpuscles 111, 112
cosmic 67
cosmic rays 67
cosmos 121, 122, 126, 128
counter example xix
critical 39, 26, 26
critical pressure 26
cycle 55
cycloid 83, 87
cycloidal 71

D

Davisson
D.B. , 167, 168
debris 126, 129
de Broglie 154
deBroglie 79, 81, 132, 136, 139,
    140, 142, 144, 145
decay 34, 152, 76, 58, 56, 51, 56
demise 128, 129
diffraction
dinosaur 129
dinosaurs 128, 129
disintegration 126, 127
displacement 6, 54
DNA xvi
Doppler 119
doublet 40
Douglas Corporation 115
Dr. Leon A. viii
dumbbell 44, 45
dynamic pressure 38

E

earth 5, 153, 117, 121, 126, 129,
    153, 55
Earth's absolute velocity 153
eccentricity 72, 73, 78
economic 151
egg 55
Einstein viii, 69, 153, 84, 153
electromagnetic coupling con-
    stant 20
electromagnetic force 51, 53
electromagnetic force velocity 61
electron 53, 58
electrostatic ix, xix, 64, 65, 40,
    43, 44, 45, 47, 49, 60, 81, 82,
    84, 151, 152, 153, 154, 53,
    52, 53, 52
electrostatic field 43, 45, 49, 60,

81, 151, 152, 153
electrostatic force 40, 44, 60, 61,
    81, 82, 152, 84, 86, 154
electrostatics viii, 44, 152
elementary matter particles 52
ellipse 72, 73, 153
elliptic 72, 73, 75, 83
elliptical 61
Energy xii, xiii, 130, 160, 162,
    164
entropy 149, 55
expanding universe 115, 116, 117
explosion 126, 127, 128
explosions 55

F

fifth force 51
fine structure viii, 49, 61, 154
fine structure constant viii, 61,
    115, 154
fission 127
Five Forces xi, 51
fluid sink 149, 25
foci 72
forced 152
Ford
free molecular flow 21
fusion 123, 124
future

G

gaseous ether 68
Gaussian 7
geometry 149
Germer
getting xvi
God
Graham Wells viii

gravitation 5, 64, 46, 152, 152, 154, 157, 51
gravitational field 64, , 123, 124, 53
gravity vii, viii, 79
group vii, viii, 70, 115, 56
groups xv, 22

H

half life 128
Halliday
hard 3, 6, 43, 58
Harmon viii, 63, 71, 115, 168
Higgs
Hill
homogeneous 6, 8, 55
Hopper ix
Hubble , 113, 116, 117, 119
Hydrodynamics , 167
hydrogen 65, 60, 63, , 109, 152, 112, 153, 118, 119, 123, 124, 125, 126

I

ideal , 26
induction
interference 111, 112

J

J.M. , 167, 168
Johnson vii

K

Kennard 7
kinetic energy 123
knowledge 150

L

law xix, 149
laws xv
Lee xvii, 56
life 127, 128
linear momentum 69, 78, 82, 113
lived
loop 47, 49, 152, 152

M

magnetic moment 151
magnetism xix
mass growth viii, 71, 153, 79, 80
matter shortening 80
Maxwell viii, 7, 8, 149, 150, 55
Maxwell-Boltzmann viii, 7, 8, 149, 113, 150, 55
Maxwell's
McClusky 75, 168
McDonnell vii, viii, 115, 168
McDonnell Douglas vii, viii, 115, 168
McDonnell Douglas Corporation 115
McRae ix
mean free path xx, 4, 6, 44, 65, 44, 45, 150, , 68, 158, 115, 122, 151, 21, 23
mean speed 8, 61, 152, 115, 20
meteorite 129
microrocket 21
mirror 54, 56, 58

Mississippi State University viii
Mohr , 167
mRNA xvi
muon 122

N

neuron 126
neutron , 152, 123, 124, 126, 152
neutron star 124, 126, 127
Newton 111, 112
Newtonian viii, 43, 153, 80, 56
nozzle 25
nozzles 25
numerological viii

O

organisms 129

P

paradigm 137
Patterson vii, ix
photon mass 116, 119
planar coil 71, 83
Planck 68, 79, 114, 116, 157
plane spiral path 153
plane spiral paths 51
plus 73
postulates 3, 6, 63
pump 150

Q

quantum electrodynamics 154,
    20
quantum mechanics xvii, xix, 1,
    137, 148

R

radiation 3, 4, 68, 113, 76
radius of the observable universe
    119, 120
R.C. , 167
red shift 113, 116
reference 47, 72, 81, 85, 131
Reference xiii
reference frame 72, 81
regenerated 129
relativity viii, 153, 76, 79, 80, 83,
    85
reverse aging 54, 55
rms speed 8, 115
Robert M. vii

S

Schrödinger xiii, xvii, xix, 1, 131,
    132, 133, 154, 136, 138, 139,
    142, 143, 144, 145, 147, 148,
    154
second law of thermodynamics
    xix, 149
set 44, 66, 117, 149
sink , 38, 39, 40, 149, 150, 43, 20,
    21, 23, 25, 25
sonic sphere , 150, 39, 41, 20, 21,
    20, 23, 22
source , 149, 119, 150, 51
special relativity 80
special theory of relativity viii,

76, 85
specific heat 26
speed of sound 26
spin 45, 47, 56, 58
spiritual 129
stable 8, 150, 151, 153, 23, 23
star 117, 120, 123, 124, 126, 127
stars xv, 5, 110, 113, 121, 123,
        126, 55
Steinert viii, ix, 46
Stokes ix
strain xvi
string 73, 109, 110, 116
strong nuclear force 34, 37, 40,
        41, 42, 43, 61, 152, 52, 52
subsonic 20, 21
superconductivity 113
S.W. , 168
Symmetry xi, 54

T

tautology
Taylor , 167
TCP theorem 54
temperature 113
theory of relativity viii, 153, 76,
        85
think 35, 121, 160
thinking viii
Thomas 111, 167
thrust 33, 150, 34, 151, 47, 48, 51,
        51, 51, 51
time dilation 76, 80
tired light 113, 116
torque
torsion xvi
torsional xvi
torsional strain xvi

U

unified field theory 66
United States iii
universe vii, xv, xix, 1, 3, 4, 6, 8,
        43, 67, 80, 81, 63, , 68, 84,
        85, 149, 150, 110, 113, 115,
        116, 117, 119, 120, 22, 51,
        55
usable energy 150, 22, 51

V

vacuum 150, 150, 25
velocity of the earth

W

weak nuclear force 152, 52, 51
Weeden ix
Wells viii
Whittaker 43
Wood vii, 63, 115, 168

Y

Yeatman viii
Young 111

Z

zero 40, 41, 41, 42, 82, 21, 25, 84,
        86, 55